WEIRD AND UNUSUAL TRIVIA

Publications International, Ltd.

ISBN: 978-1-64558-015-7

Manufactured in China.

8 7 6 5 4 3 2 1

CONTENTS

Black Sheep • Going To War in Style • A Wing and a Prayer • The Hundred Years' War—in Five Minutes • Ethan Allen • How Old Is Old Ironsides? • U.S. Veterans of the Revolutionary War • Doin' the Duck and Cover • The Toledo War • Mutiny or Mistake? • Maass-ively Brave

CHAPTER 1

MATTERS OF LIFE AND DEATH

9 Strange Last Wills and Testaments

A will is supposed to help surviving family and friends dispose of your estate after you've passed away. Many people use it as an opportunity to send a message from beyond the grave, either by punishing potential heirs with nothing or perhaps by giving away something fun or unusual to remember them by. Where there's a will, there's a way, so make sure you have a good will before you go away for good.

1. Harry Houdini: Harry Houdini, born in 1874, was considered the greatest magician and escape artist of his era. When he died in 1926 from a ruptured appendix, Houdini left his magician's equipment to his brother Theodore, his former partner who performed under the name Hardeen. His library of books on magic and the occult was offered to the American Society for Psychical Research on the condition that J. Malcolm Bird, research officer and editor of the *ASPR Journal*, resign. When Bird refused, the collection went instead to the Library of Congress. The rabbits he pulled out of his hat went to the children of friends. Houdini left his wife a secret code—ten words chosen at random—that he would use to contact her from the afterlife. His wife held annual séances on Halloween for ten years after his death, but Houdini never appeared.

2. Marie Curie: Born in Russian-occupied Poland in 1867, Marie Curie moved to Paris at age 24 to study science. As a physicist and chemist, Madame Curie was a pioneer in the early field of radioactivity, later becoming the first two-time Nobel laureate and the only person to win Nobel Prizes in two different fields of science—physics and chemistry. When she died in 1934, a gram of pure radium, originally received as a gift from the women of America, was her only property of substantial worth. Her will stated: "The value of the element being too great to transfer to a personal heritage, I desire to will the gram of radium to the University of Paris on the condition that my daughter, Irene Curie, shall have entire liberty to use this gram . . . according to the conditions under which her scientific researches shall be pursued." Element 96, Curium (Cm), was named in honor of Marie and her husband, Pierre.

3. William Randolph Hearst: Multimillionaire newspaper magnate William Randolph Hearst was born in San Francisco in 1863. When he died in 1951, in accordance with his will, his $59.5 million estate was divided into three trusts—one each for his widow, sons, and the Hearst Foundation for Charitable Purposes. Challenging those who claimed he had children out of wedlock, Hearst willed anyone who could prove "that he or she is a child of mine . . . the sum of one dollar. I hereby declare that any such asserted claim . . . would be utterly false." No one claimed it. The book-length will included the disposition of his $30 million castle near San Simeon, California. The University of California could have had it but decided it was too expensive to maintain, so the state government took it, and it is now a state and national historic landmark open for public tours.

4. Jonathan Jackson: Animal lover Jonathan Jackson died around 1880. His will stipulated that: "It is man's duty as lord of animals to watch over and protect the lesser and feebler." So, he left money for the creation of a cat house—a place where cats

could enjoy comforts such as bedrooms, a dining hall, an auditorium to listen to live accordion music, an exercise room, and a specially designed roof for climbing without risking any of their nine lives.

5. S. Sanborn: When S. Sanborn, an American hatmaker, died in 1871, he left his body to science, bequeathing it to Oliver Wendell Holmes, Sr., (then a professor of anatomy at Harvard Medical School) and one of Holmes's colleagues. The will stipulated that two drums were to be made out of Sanborn's skin and given to a friend on the condition that every June 17 at dawn he would pound out the tune "Yankee Doodle" at Bunker Hill to commemorate the anniversary of the famous Revolutionary War battle. The rest of his body was "to be composted for a fertilizer to contribute to the growth of an American elm, to be planted in some rural thoroughfare."

6. John Bowman: Vermont tanner John Bowman believed that after his death, he, his dead wife, and two daughters would be reincarnated together. When he died in 1891, his will provided a $50,000 trust fund for the maintenance of his 21-room mansion and mausoleum. The will required servants to serve dinner every night just in case the Bowmans were hungry when they returned from the dead. This stipulation was carried out until 1950, when the trust money ran out.

7. James Kidd: James Kidd, an Arizona hermit and miner, disappeared in 1949 and was legally declared dead in 1956. His handwritten will was found in 1963 and stipulated that his $275,000 estate should "go in a research for some scientific proof of a soul of a human body which leaves at death." More than 100 petitions for the inheritance were dismissed by the court. In 1971, the money was awarded to the American Society for Psychical Research in New York City, although it failed to prove the soul's existence.

MATTERS OF LIFE AND DEATH

8. Eleanor E. Ritchey: Eleanor E. Ritchey, heiress to the Quaker State Refining Corporation, passed on her $4.5 million fortune to her 150 dogs when she died in Florida in 1968. The will was contested, and in 1973 the dogs received $9 million. By the time the estate was finally settled its value had jumped to $14 million but only 73 of the dogs were still alive. When the last dog died in 1984, the remainder of the estate went to the Auburn University Research Foundation for research into animal diseases.

9. Janis Joplin: Janis Joplin was born in Texas in 1943. In her brief career as a rock and blues singer, she recorded four albums containing a number of rock classics, including "Piece of My Heart," "To Love Somebody," and "Me and Bobby McGee." Known for her heavy drinking and drug use, she died of an overdose on October 4, 1970. Janis made changes to her will just two days before her death. She set aside $2,500 to pay for a posthumous all-night party for 200 guests at her favorite pub in San Anselmo, California, "so my friends can get blasted after I'm gone." The bulk of her estate reportedly went to her parents.

Portrait of a Killer

Though some argue that humans are the most dangerous creatures on Earth, the distinction actually belongs to a tiny, common insect.

Diseases transmitted via mosquito bites have caused more death and misery than the total number of casualties and deaths suffered in all of history's wars. The Roman Empire crumbled in the 3rd and 4th centuries when the Legions,

 decimated by malaria, were unable to repel barbarian invaders. American planners took control of the Panama Canal construction project when French interests withdrew after losing more than 22,000 workers to mosquito-borne illnesses over an eight-year period.

More than 2,500 species of mosquito spread disease throughout the world. The killers begin life as larvae, hatched in almost any kind of water. Within one week, adults emerge to ply their deadly trade.

As with other species, the female is the deadlier of the two sexes, while the male concerns himself with nothing more than fertilizing eggs. Females need blood to nourish their eggs; unfortunately, the process of collecting this food supply can wreak havoc on humans.

Mosquitoes can fly more than 20 miles from the water source in which they were born. Sensory glands allow the insects to detect carbon monoxide exhaled by their victims and lactic acid found in perspiration. When a female mosquito dips her proboscis into an unwilling victim, she transfers microorganisms through her saliva into her donor. These are responsible for some of the world's most deadly and debilitating diseases, which include malaria; Yellow, Dengue, and Rift Valley fevers; West Nile virus; and at least six different forms of encephalitis.

Howard Hughes: The Paragon of Paranoia

The sad condition of reclusive mogul Howard Hughes in his last years is well known—he became a bearded, emaciated, germaphobe hiding in his Las Vegas hotel room. But the actual details of Hughes's strange life are even more shocking.

A Golden Youth

Born in Houston, Texas, in 1905, to an overprotective mother and an entrepreneur father who made a fortune from inventing a special drill bit, Howard Hughes moved to California shortly after his mother died when he was 17. There he was exposed to the Hollywood film industry through his screenwriter uncle, Rupert Hughes. When his father died two years later, Howard inherited nearly a million dollars while still in his teens. Through shrewd investments he managed to parlay that into a serious fortune, which gave him the means to pursue his interest in films that had been ignited by Uncle Rupert.

Hughes became the darling of Hollywood beauty queens. He produced several successful films, including *Scarface and Hell's Angels* (which cost him nearly $4 million of his own money).

Next, he turned his imaginative talents to aviation. He formed his own aircraft company, built many planes himself, and broke a variety of world records. He even won a Congressional Gold Medal in 1939 for his achievements in aviation. But his life took a drastic turn in 1946, when he suffered injuries in a plane crash that led to a lifelong addiction to painkillers. He also downed several seaplanes and was involved in a few auto accidents—and perhaps these mishaps helped give him the idea that the world was out to get him.

Evolving into a Hermit

Hughes married movie starlet Jean Peters in 1957, but they spent little time together and later divorced. He eventually moved to Las Vegas and bought the Desert Inn so he could turn its penthouse into his personal safe house.

In 1968, *Fortune* magazine named Howard Hughes the richest man in America, with an estimated wealth of $1 billion. But his personal eccentricities mounted almost as quickly as his for-tune. Hughes was afraid of outside contamination of all sorts, from unseen bacteria to city water systems to entire ethnic groups. He was even afraid of children. He also burned his clothes if he found out that someone he knew had an illness.

Dangerously Decrepit

Biographer Michael Drosnin provided shocking details about Hughes's lifestyle: He spent most of his time naked, his mat-ted hair hanging shoulder-length and his nails so long they curled over. Hughes even stored his own urine in glass jars. His home was choked with old newspapers and used tissues, and he sometimes wore empty tissue boxes as shoes. Hughes was obsessive, however, about organizing the memos that he scribbled on hundreds of yellow pads.

Hughes usually ate just one meal per day, and he sometimes subsisted on dessert alone. His dental hygiene was abysmal—his teeth literally rotted in his mouth. When he finally died from heart failure in 1976, the formerly robust man weighed only 90 pounds. He had grown too weak to handle his codeine syringe and had turned himself into a human pincushion, with five broken hypodermic needles embedded in his arms.

He Couldn't Take It with Him

Howard Hughes died without a will, although he had kept his aides in line for nearly two decades by dangling the promise of a fat windfall upon his death. In the end, his giant estate was inherited by some cousins.

Hughes left another legacy—his long list of achievements. He built the first communication satellite—the kind used today to link the far corners of the world. His Hughes Aircraft

Company greatly advanced modern aviation. And he produced award-winning movies. However, the full extent of his influence on the world will probably never be known.

Burn, Baby, Burn

For the last few decades, late-night comedians have joked that Cleveland is the only city in the world where both the mayor's hair and the river have caught fire.

Around noon on a sleepy Sunday in June 1969, a train car was making its way across the Cuyahoga River in downtown Cleveland. A spark from a broken wheel touched off an explosion in a pool of volatile sludge in the water below. The river, which moved slowly through the city center, tended to collect logs, railroad ties, and other debris around the foot of the trestles. These provided additional fuel, while the water itself was full of petroleum and other industrial runoff from steel plants upstream.

The blaze quickly reached heights of five stories and consumed the Norfolk & Western Railway Company trestle. Three fire departments sent teams to battle the flames from shore, while a fireboat worked from the water. The fire was out within 20 minutes, before any members of the press could even make it to the site to snap a picture. The Norfolk & Western trestle incurred $45,000 worth of damage and had to be closed and rebuilt. A nearby trestle used by the Newburgh & South Shore Railway only needed $5,000 worth of repairs. There were no deaths attributed to the fire.

The next day, the newspapers published photos from the aftermath of the blaze, merely smoldering embers and trestle rails bent from the heat. As river fires went—even in Cleveland's history—the damage wasn't bad, but the fallout from the fire is still felt today.

The Crooked River's Crooked Past

The Cuyahoga River—which means "Crooked River" in the Iroquois language—has been an important waterway since European fur traders moved into the area in the 1700s. As Ohio's population increased and the state became a center for the steel and coal industries, the river assumed a vital role in Ohio's industrial landscape. Factories popped up along its length, especially in the stretch between Akron and Cleveland.

From the start of the Industrial Revolution through the mid-1900s, the river's pollution (which was notable even then) wasn't necessarily viewed as problematic. Instead, the river's filth was a sign of industrial progress, success, and wealth.

The infamous fire of 1969 was not the first—or even the most destructive—river fire in Cleveland's history. The first recorded Cuyahoga River fire was in 1868 and was followed by successive fires in 1883, 1887, 1912, 1922, 1936, 1941, 1948, and 1952. The 1912 blaze was responsible for five deaths, as the fire spread through the shipyards before being contained. The 1952 fire caused $1.5 million worth of damage and yielded dramatic photographs that, even now, erroneously accompany any discussion of the unphotographed 1969 fire.

The Right Fire at the Right Time

Perhaps the fire of 1969 is still discussed because of two key factors: *Time* magazine and the shifting attitudes of the public.

Two months after the fire, *Time* published a major article about it—giving it more column inches (and more eyeballs) than the local newspapers had. Aside from the lurid descriptions of the putrid Cuyahoga ("Some River! Chocolate-brown, oily, bubbling with subsurface gases, it oozes rather than flows"), the article was accompanied by dramatic photographs of the destructive 1952 fire.

No matter how persuasive the article was, it would not have made a difference if people hadn't been ready to hear its message. By 1969, there was a growing ecology movement, and the general public was becoming concerned about lasting damage being done to the country's waterways. The *Time* article encapsulated these fears, complete with horrifying photographs, and Cleveland—specifically the Cuyahoga—became the symbol for industrial pollution.

Thus, the fire of 1969 became a turning point for the environmental movement. In 1972, Congress passed the Federal Water Pollution Control Amendments, also known as the Clean Water Act. This began to regulate the disposal of industrial waste and by-products, and it required cities to improve sewage treatment plants and processes and monitor the storm water and overflow.

Despite all this, the state of the river did not immediately improve. In a study conducted on the length of river between Akron and Cleveland in 1972, marine biologists could not find a single living fish or even any samples of species known to thrive in polluted water. By 1994, however, fish-eating birds such as the great blue heron and bald eagle began returning to the region, and in 2000, the Ohio Environmental Protection Agency reported finding 62 species of fish, including clean water species such as steelhead trout.

After Burn

More than 40 years after the fire-heard-'round-the-world, the Cuyahoga could be considered a "clean" river. Even the much-maligned stretch between Cleveland and Akron meets or exceeds most of the requirements set forth in the Clean Water Act. But the Cuyahoga's filthy reputation lives on, in standard fodder for comedy club routines, in Randy Newman's song "Burn On," even in the Great Lakes Brewing Company's tasty Burning River Pale Ale—and certainly in Cleveland's collective memory.

A local ribbon-cutting mishap turned to embarrassment on a national level when Cleveland mayor Ralph Perk accidentally set his hair on fire on October 16, 1972. The ceremony was for a welding convention, the ribbon was a strip of metal, and Perk was doing the cutting with an acetylene torch.

Life After Death:
Decapitation Doesn't Always Mark the End

A person can't remain conscious long enough after being beheaded to plan and exact revenge on the executioners, but it seems that a severed noggin can get in a final thought or two.

Strange Stories of Beheadings

There are many fantastic stories of living, angry heads from the heyday of decapitation. Charlotte Corday, who was executed in 1793 for the assassination of French radical leader Jean-Paul Marat, reportedly blushed when the executioner slapped her severed head. The heads of two rivals in the French National Assembly allegedly spent their last seconds biting each other.

And legend has it that when the executioner held aloft the heart of just-decapitated Sir Everard Digby, a conspirator in the Gunpowder Plot of 1605, and said, "Here lies the heart of a traitor," Digby's head mouthed, "Thou liest."

These are extremely tall tales, but more recent accounts are fairly credible. The most famous is from a French physician named Dr. Beaurieux, who witnessed the execution of a criminal named Languille in 1905. Beaurieux wrote:

"The face relaxed, the lids half closed on the eyeballs, leaving only the white of the conjunctiva visible. It was then that I called in a strong, sharp voice: 'Languille!' I saw the eyelids slowly lift up, without any spasmodic contractions—I insist advisedly on this peculiarity—but with an even movement, quite distinct and normal, such as happens in everyday life, with people awakened or torn from their thoughts. Next Languille's eyes very definitely fixed themselves on mine and the pupils focused themselves. I was not, then, dealing with the sort of vague dull look without any expression, that can be observed any day in dying people to whom one speaks: I was dealing with undeniably living eyes which were looking at me."

I'm Not Dead Yet

As horrific as the possibility seems, it is biologically feasible to temporarily survive a decapitation. The brain can still function as long as it receives oxygen delivered via blood. While the trauma of the final cut and sudden drop in blood pressure would likely cause fainting, there still would be enough blood available to make consciousness possible. Exactly how much consciousness isn't clear, but the likely cap is about 15 seconds.

The next logical question is, what might the beheaded be thinking in these final seconds? Here's a possibility: "Ouch!"

The Necessities of Survival

When rock climber Aron Ralston's arm was pinned by a boulder while he was out on a climbing trip, he did the unthinkable in order to survive.

In 2003, Aron Ralston went on a hiking trip in Utah's Blue John Canyon. Feeling confident and adventurous, Ralston, an experienced mountain climber, set out on the solo trip, neglecting to tell anyone he was leaving. The trip began perfectly: nice hike, beautiful day. Then, as Ralston tried to negotiate a narrow opening in the canyon, an 800-pound boulder fell and pinned his right forearm, completely crushing it.

It looked like there would be no escape. Ralston was completely trapped under the boulder, and his hand and arm were deadened due to the pressure of the blow. For the next five days, Ralston concentrated on staying alive, warding off exhaustion, hypothermia, and dehydration. He knew no one would be looking for him, since no one knew of his trip. Assuming the worst, he carved his name into the rock that held him down, along with what he thought would be his death date. Using a video camera he had packed with his supplies, he taped his goodbyes to his friends and family.

On the sixth day of being trapped, delirious and starving, Ralston made a hellish decision: He would cut off his own arm to escape. Bracing his arm against a climbing tool called a chockstone, Ralston snapped both his radius and ulna bones and applied a tourniquet with some rags he had on hand. Using the knife blade of his multi-tool, he then cut through the soft tissue around the broken bones and tore through tendons with the tool's pliers. The makeshift operation took about an hour.

After he was loose, Ralston still had to rappel down a 65-foot-tall cliff, then hike eight miles to his parked truck. Dehydrated and badly injured, he walked to the nearest trail and was finally discovered. A helicopter team flew Ralston to the nearest hospital where he was stabilized and sent immediately into surgery to clean up and protect what was left of his arm. He later received a prosthetic limb.

Once the press caught wind of his story, Ralston became a celebrity. These days, Ralston works as a motivational speaker, and he still goes on climbing trips, regularly setting new records. In 2010, his story was made into a movie called *127 Hours*.

Arlington National Cemetery

Arlington is much more than just a cemetery—it's a tribute to those who fought for freedom. Here are nine intriguing facts you probably didn't know.

1. The land was once owned by George Washington Parke Custis, who was the adopted grandson of President George Washington. The land, and the mansion built on it, passed via inheritance and marriage to Confederate General Robert E. Lee.

2. The property was confiscated by the U.S. government during the Civil War. The 200-acre tract was officially designated a military cemetery on June 15, 1864, by Secretary of War Edwin Stanton.

3. More than 300,000 people are buried there. They represent all of the nation's wars, from the Revolutionary War to the wars in Iraq and Afghanistan.

4. Those who died before the Civil War were reinterred after 1900.

5. Arlington National Cemetery boasts the second largest number of burials (for a national cemetery in the United States), after Calverton National Cemetery near Riverhead, New York.

6. An average of 28 funerals per day are conducted at Arlington National Cemetery. The flags on the cemetery grounds are flown at half-mast from a half-hour before the first funeral of the day until a half-hour after the final funeral ends. Funerals are not usually held on weekends.

7. The remains of an unknown serviceman from the Vietnam War, interred in 1984, were disinterred in 1998 after being positively identified as Air Force 1st Lt. Michael J. Blassie.

8. The Tomb of the Unknowns is guarded 24 hours a day by members of the 3rd U.S. Infantry, also known as "The Old Guard." They began guarding the tomb on April 6, 1948.

9. Among the famous people buried in Arlington National Cemetery are presidents William Howard Taft and John F. Kennedy, civil rights activist Medgar Evers, mystery writer Dashiell Hammett, Supreme Court justice Thurgood Marshall, arctic explorers Robert Peary and Matthew Henson, and band leader Glenn Miller.

Death—Isn't It Ironic?

No matter who you are, it's inevitable: Your time on this earth will end. But some people have a way of shuffling off this mortal coil with a bit more ironic poignancy.

- In 1936, a picture of baby George Story was featured in the first issue of *Life* magazine. Story died in 2000 at age 63, just after the magazine announced it would be shutting down. *Life* carried an article about his death from heart failure in its final issue.

- In the early 1960s, Ken Hubbs was a Gold Glove second baseman for the Chicago Cubs. The young standout had a lifelong fear of flying, so to overcome it, he decided to take flying lessons. In 1964, shortly after earning his pilot's license, Hubbs was killed when his plane went down during a snowstorm.

- While defending an accused murderer in 1871, attorney Clement Vallandigham argued that the victim accidentally killed himself as he tried to draw his pistol. Demonstrating his theory for the court, the lawyer fatally shot himself in the process. The jury acquitted his client and Vallandigham won the case posthumously.

- Private detective Allan Pinkerton built his career on secrecy and his ability to keep his mouth shut. However, biting his tongue literally killed him when he tripped while out for a walk, severely cutting his tongue. The injury became infected and led to his death in 1884.

- When he appeared on *The Dick Cavett Show* in 1971, writer and healthy living advocate Jerome I. Rodale claimed, "I've decided to live to be a hundred," and "I never felt better in my life!" Moments later, still in his seat on stage, the 72-year-old Rodale died of a heart attack. The episode never aired.

- South Korean Lee Seung Seop loved playing video games more than anything. His obsession caused him to lose his job and his girlfriend and eventually took his life as well. In August

2005, after playing a video game at an Internet cafe for 50 consecutive hours, he died at age 28 from dehydration, exhaustion, and heart failure.

• Jim Fixx advocated running as a cure-all, helping develop the fitness craze of the late twentieth century. However, in 1984, he died from a heart attack while jogging. Autopsy results showed he suffered from severely clogged arteries.

• Thomas Midgley Jr. was a brilliant engineer and inventor who held 170 patents. After contracting polio at age 51, he turned his attention to inventing a system of pulleys to help him move around in bed. In 1944, he was found dead, strangled by the pulley system that he had invented.

• At least two of the Marlboro Men—the chiseled icons of the cigarette culture—have died from lung cancer. David McLean developed emphysema in 1985 and died from lung cancer a decade later. Wayne McLaren portrayed the character in the 1970s, and although he was an antismoking advocate later in life, he still contracted cancer that spread from his lungs to his brain. He died in 1992.

• Author Olivia Goldsmith wrote *The First Wives Club*, a book that became an icon for older women whose husbands had tossed them aside for younger trophy wives. A generation of women embraced their wrinkles and weren't afraid to let the world know. Ironically, Goldsmith died while undergoing cosmetic surgery.

• Shortly before he died in a high-speed car crash, James Dean filmed a television spot promoting his new film *Giant*. The interviewer asked Dean if he had any advice for young people. "Take it easy driving," he replied. "The life you save might be mine."

Real Vampires

Can corpses prey on people from beyond the grave? In the cases outlined below, someone clearly thought they could.

In 1857, the *Wooster Republican* featured an article entitled "An Extraordinary Superstition: A corpse exhumed." According to the article, residents of Euphrata, Pennsylvania, had been shocked to hear that the body of one Sophia Bauman had been exhumed, nine years after her death, so that the body could be destroyed in order to save the lives of her surviving family members. Residents suspected that her body was "feeding" on living flesh from beyond the grave.

A Misunderstood Disease

Consumption—as tuberculosis was known in those days—was the single most common cause of death in the nineteenth century, and one of the least understood diseases of them all. When someone contracted it, it did look as though the life was slowly being sucked out of them. That the life was being sucked away by a person already dead was not exactly "superstition" so much as a form of folk medicine; in those days before contagious disease was really understood, the idea that a dead body could "feed" from the living seemed reasonable enough, and when entire families were dying of the disease, survivors became desperate. Sometimes it seemed that stopping a corpse from "feeding" on the living was the only way to halt the spread of the disease before it claimed an entire family. Two of Sophia Bauman's sisters had died of consumption, along with her mother and two of her bothers, and the family was getting desperate for a treatment.

It was believed, or at least suggested, that if the "winding sheet" got into a corpse's mouth, the dead body might "suck" on it and the "continual suction" from beneath the ground could draw life out of the living by causing consumption. Evidently someone wondered if perhaps that had happened to Sophia, and so, one Sunday morning, her coffin was exhumed from the cold ground. However, no winding sheet was in the skeleton's mouth—any such sheet would have long ago disintegrated after nine years in the grave.

Suspicions and Remedies

Around the time of Sophia's death, a similar case occurred elsewhere in New England in which a physician opened the graves of two people whose family had been ravaged by consumption. In the coffins, he found that mysterious snow white vines had grown all over the corpses, and seemed still to be growing. While the physician stated that he had no faith in an old superstition which held that plucking such a vine off the corpse would truly cure consumption in a family, he chose to cut away the vines anyway.

The belief that dead people could suck the life out of the living was not a new one at the time; it appears to have been going around for centuries, following European immigrants into the New World. But in some cases, simply removing a vine or winding sheet wasn't thought to be enough. In some, perhaps more common, versions of the practice, the method said to arrest the spread of the disease actually involved mutilating or destroying the corpse altogether.

A Grisly Discovery

In 1990, a group of children playing in Connecticut stumbled onto a long abandoned burial ground near a gravel mine. Police initially believed it was a serial killer's dumping grounds, but further inspection found that the graves were actually part of a colonial cemetery in which residents had been buried in simple wood coffins, unadorned by jewelry, with their arms at their sides or across their chest.But two of the coffins were in stone crypts, and one of them was in a red coffin with the letters *J.B.* spelled out in tacks on the outside. The bones inside had been rearranged inside of the coffin, with the skull and thigh bones set into a perfect skull and crossbones pattern. Analysis later showed that the body had been beheaded, and the bones moved around, some five years after "JB" had died. It seemed that someone had suspected JB of infecting the living from inside of his coffin.

Perhaps the exhumation was done quietly, but for these things to happen in large public ceremonies was not unknown. In 1793, hundreds of people went to a blacksmith's forge in Manchester, Vermont, to watch the heart of a suspected "vampire" be burned up in attempt to cure the dead man's wife's illness. That the whole town came out to see JB's bones be rearranged is not impossible.

The Case of Mercy Brown

Though JB's is the only such body to be studied firsthand by modern scientists, written testimonies and reports suggest that simply burning the heart and lungs seems to have been the most common way to "cure" a corpse that was thought to be feeding off the living, and this seems to have been the case with the most famous, and probably most recent "vampire," Mercy Lena Brown of Exeter, Rhode Island.

Mercy died of consumption in 1892, a year when the disease was devastating the families who had remained in Exeter, a community whose population had been cut in half over the course of a couple of generations.

By then, it was known in the medical community that consumption was caused by bacteria, not by the dead, but such scientific news spread only slowly to isolated rural towns. By the time Mercy died, two of her family members had already succumbed to the disease, and her brother was battling it too.

Desperate to save Mercy's brother when he took a turn for the worse, their father agreed to have Mercy's corpse exhumed a few months after her death to see if there was fresh blood in the heart, which was said to be a sure sign that the body was still being kept alive by feeding on living tissue. Mercy's mother and sister, who had also died of consumption, were exhumed first, though they were merely skeletons by then. The winter weather, though, had kept Mercy's own body from decomposing too much. When blood was found in her heart, it was burned on a nearby rock, and the ashes were fed to her brother.

The "cure" didn't work; Mercy's brother was dead within two months. News of the event circulated through the national press, and other cases were reported. An 1893 *Chicago Tribune* article referred to it as a "grewsome (sic) superstition" that still survived in Pennsylvania, though there's some evidence of it having occurred in Chicago itself a couple of decades before.

Mercy, though, is the last known person to have been suspected of "vampirism," but it's generally believed that only a tiny fraction of cases are still known today, as most were never written down or reported at all. There may have been hundreds more. New stories are being found in old newspapers, ancient diaries, and faded letters all the time.

A Love Eternal

Carl Tanzler couldn't have his dream girl in life, so he bided his time. When she passed away from natural causes, the door to romance—creepy, morbid, romance—suddenly sprang open.

German immigrant Carl Tanzler (AKA Count Carl Von Cosel, 1877-1952) loved to tell tall tales. While working as an X-ray technician at the U.S. Marine Hospital in Key West during the Depression years, Tanzler claimed to be an electrical inventor, the hold-

er of multiple university degrees, even a submarine captain. None of it was true. In reality Tanzler was a lonely, conflicted man caught up in a fantasy world of his own design. But this profound sense of loneliness ebbed the instant that he met hospital patient Elena Hoyos, a fetching 22-year-old Cuban woman suffering from tuberculosis. He believed that it was his destiny to be with her.

Snake Oil Therapy

Try though he might, Tanzler was mostly rebuffed by his new-found object of affection. Undeterred by her rejection, Tanzler set his sights on working to cure her and tried his best to win the approval of her family. His efforts were in vain.

After reviewing Hoyos's X-rays, Tanzler realized that she wasn't long for the world. Nevertheless, he convinced her that he could cure her with a combo of X-rays and daily doses of his special tonic, a bizarre mixture comprised of gold and water. Not surprisingly his "cure" failed as badly as his attempts at winning her hand, and Hoyos succumbed to her illness.

A Love Realized

After Hoyos's passing, Tanzler was devastated but strangely optimistic about their future together. This was a man with a plan—a depraved and twisted plan to be sure—but a plan nonetheless. By hook or by crook Hoyos would soon be his.

Hoyos was buried in a common grave. Unhappy with this arrangement, Tanzler asked Hoyos's family for permission to rebury her in a stone mausoleum. After getting the go-ahead, Tanzler discovered that Elena's body hadn't been embalmed. He hired a mortician to do the job, and her body was moved to its new home. Now Hoyos would ride out eternity with the dignity that Tanzler felt she deserved. Now, too, he could make unauthorized visits to her remains compliments of a solitary key that her parents knew nothing about.

Proving that obsessive love knows neither bounds nor boundaries, Tanzler brought her flowers and gifts each night and counted the hours in between their visits. It was said that he even installed a telephone in her tomb in hopes that she'd communicate with him. After a two-year "courtship," Tanzler removed Elena's body from the tomb and relocated it to an abandoned airplane fuselage behind the hospital.

Quality Time

Using wax, plaster of Paris and glass eyes, Tanzler restored "life" to his beloved. A shocker came when Tanzler learned that the military had plans to move the old airplane. With loving concern, Tanzler relocated Hoyos once again—this time to his house, where no one could interfere. As a sign of his commitment, Tanzler dressed Hoyos in a wedding gown and slept with her each evening. He anointed his lady with body oils, chemicals and perfumes—all the better to keep the putrid smell of decomposition at bay.

Rigor Mortis Interruptus

Tanzler made regular trips into town to obtain supplies, and that's where his troubles began. People became curious when they saw Tanzler purchasing women's clothing and perfumes, especially after a paper boy claimed to have seen him dancing with a big doll through a window at his home. Before long, rumors began to circulate. Could Tanzler be spending time with the corpse of Elena Hoyos?

Elena's sister Nana asked Tanzler if there was any truth to the macabre assertions. Not wanting to disgrace himself any further, he came clean. He led Nana to his house to show her how beautifully he'd arranged Elena. There, propped up in a chair, was Nana's dead sister, looking much the worse for wear after nine long years. Nana, disbelieving, summoned the police.

Ghoul's Luck

Tanzler was arrested and charged with grave robbing and abusing a corpse. He was ultimately convicted of grave robbery, but got off on a technicality. The statute of limitations had expired and Tanzler received no jail time for his offense.

Hopeless Romantic

The case drew great publicity. For this reason, authorities decided to put Elena's corpse on public display at a funeral home. After the viewing, her remains were placed in a metal box and buried in a secret location.

Despite the setbacks, Tanzler's obsession with Elena continued. For the rest of his years he carried his beloved close to his heart—exceedingly close. When authorities discovered Tanzler dead at his house in 1952, he was clutching a life-sized doll with a face that looked uncannily like Elena's.

Sister Aimee, Radio Sensation

Did an early twentieth century evangelist stage her own death?

Sister Aimee Semple McPherson (1890–1944) was a woman far ahead of her time. In a male-driven society, McPherson founded a religious movement known as the Foursquare Church. Using her natural flamboyance and utilizing modern technologies such as radio, McPherson reached thousands with her Pentecostal message of hope, deliverance, and salvation. But turbulent waters awaited McPherson. Before the evangelist could grow her church to its fullest potential, she'd first have to survive her own "death."

The Seed Is Planted

McPherson was something of a firebrand right from the get-go. Born Aimee Elizabeth Kennedy in Salford, Ontario, the future evangelist was daughter to James Kennedy, a farmer, and Mildred "Minnie" Kennedy, a Salvation Army worker. As a teenager, the inquisitive Aimee often came to loggerheads with pastors over such weighty issues as faith and science—even as she questioned the teaching of evolution in public schools.

In 1908, Aimee married Robert James Semple, a Pentecostal missionary from Ireland. The marriage was short-lived. Semple died from malaria in 1910, but their union produced a daughter, Roberta Star Semple, born that same year.

Working as a Salvation Army employee alongside her mother, Aimee married accountant Harold Stewart McPherson in 1912. One year later they had a son, Rolf Potter Kennedy McPherson. But this marriage would also dissolve. Citing desertion as the cause for their rift, Harold McPherson divorced his wife in 1921.

By this point McPherson was well on her way as an evangelist. In 1924 she began to broadcast her sermons over the radio. This new electronic "reach"—coupled with McPherson's flair for drama—drew hordes into her fold. From an evangelistic standpoint, it was the best of times. But as Dickens said in the opening line of *A Tale of Two Cities*, such heady times rarely come without strings attached. McPherson would soon experience this directly—ostensibly from the afterworld.

Gone with the Tide?

On May 18, 1926, the shocking news broke like a wave crashing against a beach: Nationally famous evangelist Aimee Semple McPherson had gone missing while swimming in the Pacific Ocean near Venice Beach, California. She was presumed to be drowned.

Adding to the tragedy, two of her congregants perished while searching for her in the ocean. Despite continued efforts, no trace of McPherson—or her body—could be found.

From Death Comes Life

Oddly, police received hundreds of tips and leads that suggested that McPherson hadn't drowned at all. One letter— signed "The Avengers"—said that Aimee had been kidnapped and demanded $500,000 for her safe return. One month later, a very alive McPherson emerged near Douglas, Arizona. She claimed she had been kidnapped and held in a shack in Mexico. No such shack, however, could be found.

Even stranger, radio operator and church employee Kenneth G. Ormiston vanished at precisely the same time as McPherson. Gossip spread like wildfire that the married Ormiston and McPherson had in fact shacked up for a month of tawdry romance. Charges of perjury and manufacturing evidence were brought against McPherson and Ormiston but were inexplicably dropped months later.

Scandal Sells

Despite the scandal, McPherson's church continued to grow by leaps and bounds. McPherson married a third time in 1931, divorcing by 1934. In 1944, Aimee Semple McPherson died from an overdose of sedatives. Her death was ruled accidental, but many believed that McPherson had in fact committed suicide. Whatever the cause of her death, she left behind a strong legacy. By the end of the twentieth century, the church she founded boasted more than two million members worldwide.

The Catacombs of Paris

Six million souls are buried beneath the "City of Lights." According to reports, some aren't happy with the arrangement.

For the Poor No More

Indigence and indignity seem to go hand in hand. In Paris during the twelfth century, financially-strapped souls were buried in mass burial grounds, and unenviable fate worlds removed from the dignified private plots and ceremonies available to the rich.

During the late eighteenth century, however, the poor would have the last ghoulish laugh, when Parisian cemeteries were filled to overflowing and all deceased people, regardless of their station in life, were committed to a common grave. *Touché!*

An abandoned network of underground stone quarries beneath Paris was chosen for this macabre purpose. The Catacombs of Paris offered a way to sidestep the problem of decaying flesh leaching into the ground, a bona fide health concern for a society that drew its drinking water from underground wells. Over time, more than six million people were committed to the 'combs. Here, they would sleep away eternity.
Or would they?

Rest in Pieces

Piled up in a 180-mile stretch of tunnels, the dead bodies of the Catacombs are anything but ordinary. The remaining skulls and bones have been stacked in orderly if bizarre fashion to create grisly monuments and walls, and the dank, dark setting is conducive to frights both real and imagined.

This spooky underground netherworld currently operates as the Catacombs Museum. For a fee, visitors can walk through a section of the tunnels and commune with the dead.

Contrary to popular belief, visits to this no-man's-land of death and decay is anything but new. Tours of the Catacombs have taken place since 1867. Members of the French Resistance used the network of tunnels to hide out from the Germans during World War II, and the Germans used a portion of the Catacombs as a bunker during the same world-shaping conflict.

Rude Awakenings

If visitors to the Catacombs should forget where they are, a sign reading "*Arrête! C'est ici l'empire de la mort*" (Stop! This is the Empire of Death) gives fair warning about what they will soon encounter. When six million corpses are crammed together, there are bound to be a few spirits that grow restless. The

Catacombs are deemed one of the most haunted places on earth by travel journals, and reports of ghostly sightings and other paranormal encounters seem as numerous as the bodies themselves.

Unfriendly Ghosts

Based on the accounts of startled witnesses, those who expect to find friendly ghosts in the Catacombs will be terrified by the ones that they do encounter. Some visitors claim that they were "touched by unseen hands." Others tell of an uncanny feeling of being watched as they walked through the underground labyrinth. Several people tell of an ominous group of shadows that followed them step for step as they moved through the tunnels. Photos of apparitions snapped by visitors are plentiful and varied. Creepy cold spots and inexplicable photographic orbs have also been detected. Some even claim to have been choked with great might by a frightening invisible force.

Frightful Fun

Tours have occasionally been cut short when visitors grew hysterical due to such ghostly pranks, but the popularity of the Catacombs as a haven for visitors has remained strong throughout the years. In fact, more than one million visitors make the subterranean journey each year.

From the Grave

A former U.S. congressman and Virginia plantation owner was buried in a curious manner—in order to keep a watchful eye on his slaves even in death. In a day and age when slavery was the norm, Colonel George Hancock (1754–1820) was a decided cut below. When dealing with the slaves on his estate, Hancock took special pleasure in their toil and torment. In his sadistic view, he had every right. After all, weren't these the same liars that shunned their duties whenever he turned his back? Driven by such relentless suspicion, Hancock was forever on the look-out for better ways to keep his slaves in line—during and after his natural life.

Ambitious and Malicious

On paper, one would hardly find reason to suspect that Hancock was anything but a refined southern gentleman. A colonel during the Revolutionary War, Hancock became a lawyer at the war's end and eventually entered politics. He was elected as a U.S. congressman and served from 1793 to 1797.

When his congressional term ended, Hancock acquired the southern Virginia estate of Fotheringay—named for the English castle where Mary, Queen of Scots was beheaded in 1587. There, like most prosperous southern men of his day, he kept slaves.

Whenever political duties took Hancock away from his mansion, he would return with a large and bitter chip on his shoulder. So certain was he that his slaves had been misbehaving and slacking off in his absence, Hancock would dole out punishment without any proof that they had in fact done anything wrong. Completely overlooking his own sadistic ways, Hancock openly questioned the loyalty of his slaves—then punished them whether or not he had cause.

Heartbreak and Burial

In 1820, Hancock's 29-year-old daughter Julia died unexpectedly. It was more than the colonel could bear. Crushed by her sudden demise, Hancock himself took ill and passed away shortly thereafter. A double funeral service was held for father and daughter at Fotheringay. At the colonel's request, he was placed beside his daughter in a vault overlooking his acreage.

But from there, according to reports, his request got a bit strange. Apparently, it was the wish of the colonel to be buried sitting up. What was the colonel's rationale for requesting burial in an upright position? It would enable him to watch over his slaves in the fields below "and keep them from loafing on the job," he reportedly said. In death—as in life—Hancock apparently had no intention of easing up on his slaves.

Cruel Man of His Word

For years the legend of Hancock's upright burial was just that —a legend. But then a newcomer decided to delve deeper and learn the truth. In 1886, the new owner of Fotheringay, Anne Beale Edmundsun, acted. Hancock's mausoleum had begun to crumble, and she decided to enter it to learn the full extent of the damage. Of course, this would also enable her to investigate the account of his burial and separate fact from fiction.

Once inside, Edmundsun and her family members discovered a mass of bones. At the top of the heap they found a skull. They assumed the skull could only belong to Colonel Hancock, since all other family members were accounted for. Just beneath the heap were bones that appeared to be from the colonel's legs and trunk. A cruel overseer to the very end, Hancock had indeed been buried sitting up.

MATTERS OF LIFE AND DEATH

CHAPTER 2

NATURE'S WONDERS

Leave It to Beavers

They toil 365 days a year. If you knock down their work, they re-build in a matter of hours. They use just about everything they can find to forge their structures. Read on to learn more about these mad builders.

• Beavers are the largest rodents in the world after the South American capybara.

• Beavers can stay submerged in water for up to 15 minutes. Using their webbed feet for speed and their flat tail as a rudder, they can swim as fast as five mph.

• Beavers construct dams in order to deepen shallow wa-ter-ways. This in turn creates ponds, which is where beavers like to build their lodges.

• Their instinct is to stop the movement of water, which is fine for the beavers but an unnatural state of affairs for a stream. Their obsessive focus is a potentially disastrous one in streams that function as part of a community's storm water drainage system.

• Beavers are seriously resourceful. Though the majority of dams are made of mud and sticks, beaver dams have been found made of cornstalks, leaves, soybean plants,

sand, and gravel. There are even tales of dams built within city limits containing fence posts, lawn furniture, and hobbyhorses—even animal carcasses!

• One trapper found a dam made entirely of footwear. A shoe store had closed and the company apparently felt it would be a great idea to dump the leftover shoes into a nearby stream. Though this was clearly a bad idea, the beavers used the shoes to build their dam, which remained stable for years.

22 Peculiar Names for Groups of Animals

1. A shrewdness of apes
2. A battery of barracudas
3. A kaleidoscope of butterflies
4. A quiver of cobras
5. A murder of crows
6. A convocation of eagles
7. A charm of finches
8. A skulk of foxes
9. A troubling of goldfish
10. A smack of jellyfish
11. A mob of kangaroos
12. An exaltation of larks
13. A troop of monkeys
14. A parliament of owls
15. An ostentation of peacocks
16. A rookery of penguins
17. A prickle of porcupines
18. An unkindness of ravens
19. A shiver of sharks
20. A pod of whales
21. A descent of woodpeckers
22. A zeal of zebras

7 Animals that Can Be Heard for Long Distances

Animals send out messages for very specific reasons, such as to signal danger or for mating rituals. Some of these calls, like the ones that follow, are so loud they can travel through water or bounce off trees for miles to get to their recipient.

1. Blue Whale
The call of the mighty blue whale is the loudest on Earth, registering a whopping 188 decibels. (The average rock concert only reaches about 100 decibels.) Male blue whales use their deafening, rumbling call to attract mates hundreds of miles away.

2. Howler Monkey
Found in the rain forests of the Americas, this monkey grows to about four feet tall and has a howl that can travel more than two miles.

3. Elephant
When an elephant stomps its feet, the vibrations created can travel 20 miles through the ground. They receive messages through their feet, too. Research on African and Indian elephants has identified a message for warning, another for greeting, and another for announcing, "Let's go." These sounds register from 80 to 90 decibels, which is louder than most humans can yell.

4. North American Bullfrog
The name comes from the loud, deep bellow that male frogs emit. This call can be heard up to a half mile away, making them seem bigger and more ferocious than they really are. To create this resonating sound used for his mating call, the male frog pumps air back and forth between his lungs and mouth, and across his vocal cords.

5. Hyena

If you happen to hear the call of a "laughing" or spotted hyena, we recommend you leave the building. Hyenas make the staccato, high-pitched series of hee-hee-hee sounds (called "giggles" by zoologists) when they're being threatened, chased, or attacked. This disturbing "laugh" can be heard up to eight miles away.

6. African Lion

Perhaps the most recognizable animal call, the roar of a lion is used by males to chase off rivals and exhibit dominance. Female lions roar to protect their cubs and attract the attention of males. Lions have reportedly been heard roaring a whopping five miles away.

7. Northern Elephant Seal Bull

Along the coastline of California live strange-looking elephant seals, with huge snouts and big, floppy bodies. When it's time to mate, the males, or "bulls," let out a call similar to an elephant's trumpet. This call, which can be heard for several miles, lets other males—and all the females nearby—know who's in control of the area.

Hummingbirds

When early Spanish explorers first encountered hummingbirds in the New World, they called them *joyas voladoras*—or "flying jewels." But the hummingbird is more than just beautiful: Its physical capabilities put the toughest human being to shame.

• The ruby-throated hummingbird—the only hummingbird species east of Mississippi—migrates at least 2,000 miles from its breeding grounds to its wintering grounds. On the way, it crosses the Gulf of Mexico—that's 500 miles without rest. Not bad for a creature that weighs just an eighth of an ounce and is barely three inches long.

• A hovering hummingbird has an energy output per unit weight about ten times that of a person running nine miles per hour. If a person were to do the same amount of work per unit weight, he or she would expend 40 horsepower.

• A man's daily energy output is about 3,500 calories. If one were to recalculate the daily energy output of a hummingbird—eating, hovering, flying, perching, and sleeping—for a 170-pound man, it would total about 155,000 calories.

• An average man consumes about two and a half pounds of food per day. If his energy output were the same as that of a hummingbird, he would have to eat and burn off, in a single day, the equivalent of 285 pounds of hamburger, 370 pounds of potatoes, or 130 pounds of bread.

• The ruby-throated hummingbird can increase its weight by 50 percent—all of it fat—just before its winter migration. This provides extra fuel for the long, nonstop flight across the Gulf of Mexico. In comparison, a 170-pound man would have to pack on enough fat to increase his weight to 255 pounds in just a few weeks.

• The wing muscles of a hummingbird account for 25 to 30 percent of its total body weight, making it well adapted to flight. However, the hummingbird has poorly developed feet and cannot walk.

• Due to their small body size and lack of insulation, hummingbirds lose body heat rapidly. To meet their energy demands, they enter torpor (a state similar to hibernation), during which they lower their metabolic rate by about 95 percent. During torpor, the hummingbird drops its body temperature by 30º F to 40º F, and it lowers its heart rate from more than 1,200 beats per minute to as few as 50.

• Hummingbirds have the highest metabolic rate of any animal on Earth. To provide energy for flying, they must consume up to three times their body weight in food each day.

• Unlike other birds, a hummingbird can rotate its wings in a circle. It can also hover in one spot; fly up, down, sideways, and even upside down (for short distances); and is the only bird that can fly backward.

• The smallest bird on Earth is the bee hummingbird (*Calypte helenae*), native to Cuba. With a length of only two inches, the bee hummingbird can comfortably perch on the eraser of a pencil.

• The most common types of hummingbirds include the Allen's, Anna's, berylline, black-chinned, blue-throated, broad-billed, broad-tailed, buff-bellied, Costa's, Lucifer, Magnificent, ruby-throated, Rufous, violet-crowned, and white-eared.

• Like bees, hummingbirds carry pollen from one plant to another while they are feeding, thus playing an important role in plant pollination. Each bird can visit between 1,000 and 2,000 blossoms every day.

• There are about 330 different species of hummingbirds. Most of them live and remain in Central and South America, never venturing any farther north. Only 16 species of hummingbirds actually breed in North America.

NATURE'S WONDERS

Flying Fish

Find out the facts about these fascinating fish!

• To escape predators such as swordfish, tuna, and dolphins, flying fish extend their freakishly large pectoral fins as they approach the surface of the water; their velocity then launches them into the air.

• A flying fish can glide through the air for 10 to 20 feet—farther if it has a decent tailwind. It holds its outstretched pectoral fins steady and "sails" through the air, using much the same action as a flying squirrel.

• If you'd like to catch sight of a member of the Exocoetidae family, you'll have to travel to the Atlantic, Indian, or Pacific oceans, where there are more than 50 species of flying fish.

• Whiskers are not an indication that a flying fish is up there in years; it's the opposite. Young flying fish have long whiskers that sprout from the bottom jaw. These whiskers are often longer than the fish itself and disappear by adulthood.

• Attached to the eggs of the Atlantic flying fish are long, adhesive filaments that enable the eggs to affix to clumps of floating seaweed or debris for the gestation period. Without these filaments, the eggs (which are denser than water) would sink.

• Flying fish can soar high enough that sailors often find them on the decks of their ships.

How to Avoid a Mountain Lion Attack

Mountain lions, also known as cougars, pumas, and panthers, are the largest cats in North America and live in a vast area from the Yukon Territory in Canada to the Pacific Coast, the Rocky Mountains, and even Florida. Mountain lions are more plentiful than most people realize, and, though they generally avoid people, attacks do occur. These tips may help you avoid one, although no approach is guaranteed.

• Hike in groups. Mountain lions avoid crowds and noise, and the more people on the lookout, the better. If there are children in the group, make sure they are supervised.

• Be aware of your surroundings, paying attention to what's behind and above you in trees and on rocks and cliffs.

• Don't back the animal into a corner—give it a way out. It would much rather run off and survive to hunt again.

• If you encounter a mountain lion, stand still rather than try to run away. Running may cause the animal to chase you, and it's much faster than you are. Stand still while facing the mountain lion, but avoid looking it in the eye, which it takes as a sign of aggression. Watch its feet instead.

• Do things that make you appear larger and bigger than the cat, such as raising your arms over your head or holding up a jacket, a backpack, or even your mountain bike.

• Make loud noises. Growling can make you sound like something the cat would prefer not to mess with.

• Don't crouch down or bend. This makes you appear smaller and, therefore, an easy target. Don't move a lot but don't play dead. To a mountain lion, a perfectly still human looks like an entrée.

- Remain calm and don't act afraid. Like many animals, mountain lions can detect fear.

Are Bookworms Worms?

Librarians would like to banish bookworms from the stacks forever. The literature in question aren't people—they're tiny winged creatures known as book lice or barklice. They resemble flies, not worms, and they don't even like paper. They feed off mold, like in damp, mildewed tomes.

Other bugs, including silverfish and cockroaches, feast on organic substances such as the flour and cornstarch found in old library paste. Wood-boring beetles eat wood, naturally, but will consume paper made of wood pulp, too. Though beetles technically are not worms, they probably inspired the term "bookworm."

Nineteenth century French book dealer and bibliophile Étienne-Gabriel Peignot reported finding a bookworm that had burrowed clear through a set of 27 volumes, leaving a single hole, like the track of a bullet, in its wake. How far did it go? Given that many old tomes are at least three inches thick, the critter might have traveled nearly seven feet.

The best way to keep bugs out of books is to stop them before they get in. Contemporary librarians strive to maintain clean, dry buildings. Replacing wood shelving with metal discourages beetles. And those signs that say, "Please do not eat in the stacks"? Heed them. A few moldy crumbs can be a bonanza to hungry book lice. Once insects settle in, it's difficult to get rid of them. Sometimes the only effective method is professional fumigation.

Modern construction and bookbinding methods have done a lot to make libraries free of bookworms—free of the six-legged kind, that is. If you're a human bookworm, come right in. Just check your lunch at the door.

The Fate of the Passenger Pigeon

When Europeans first visited North America, the passenger pigeon was easily the most numerous bird on the continent. But by the early 1900s, it was extinct. What led to this incredible change in fortune?

Pigeons on the Wing

From the first written description of the passenger pigeon in 1534, eyewitnesses struggled with how to describe what they saw. Flights of the 16-inch-long birds were staggeringly, almost mind-numbingly big; flocks were measured in the millions, if not billions, and could be heard coming for miles. When passing overhead, a flight could block out the sun to the point that chickens would come in to roost. Passenger pigeons flew at around 60 miles an hour—one nickname dubbed the bird the "blue meteor"—but even so, a group sighted by Cotton Mather was a mile long and took hours to pass overhead. At least one explorer hesitated to detail what he had seen, for fear that the entirety of his report would be dismissed as mere exaggeration.

Settlers viewed the pigeons with trepidation. A passing flock could wreak havoc on crops, stripping fields bare and leading to famine. A flight passing overhead or roosting on your land would leave everything covered with bird droppings—a situation that would lead to more fertile soil in following years but did little to endear the creatures to farmers at that moment.

Pigeons on the Table

With such vast numbers, what could possibly have led to the extinction of the passenger pigeon? There are a number of theories, but the most likely answer seems to be the most obvious: People hunted them out of existence. Native Americans had long used the pigeons as a food source, and the Europeans followed suit, developing a systematic approach to harvesting the birds that simply outstripped their ability to reproduce. At first, the practice was an exercise in survival—a case of explorers feeding themselves on the frontier or settlers eating pigeon meat in place of the crops the birds had destroyed. However, necessity soon evolved into a matter of convenience and simple economy—the birds were cheap to put on the table.

Killing the birds in bulk was almost a trivial exercise. Initially, settlers could walk up under trees of nesting birds and simply knock them down using oars. As the birds became warier, firearms were a natural choice for hunters; flocks were so dense, one report gives a count of 132 birds blasted out of the sky with a single shot. Nets were strung across fields, easily yanking the birds from the air as they flew. Perhaps most infamously, a captive bird would be tied to a platform that was raised and then suddenly dropped; as the pigeon fluttered to the ground, other pigeons would think the decoy was alighting to feed and would fly down to join him—a practice that became the origin for the English term "stool pigeon." Hunters would catch the birds in nets, then kill them by crushing their heads between thumb and forefinger.

Pigeons on Display

By 1860, flocks had declined noticeably. By the 1890s calls went out for a moratorium on hunting the animals, but to no avail. Conservation experts tried breeding the birds in captivity to little effect; it seemed the pigeons longed for their enormous flocks and could not reproduce reliably without them.

Sightings of passenger pigeons in the wild stopped by the early 1900s. A few survivors remained in captivity, dying one by one as ornithologists looked on helplessly. The last surviving pigeon, a female named Martha, died at the Cincinnati Zoological Garden on September 1, 1914. Her body was frozen in ice and shipped to the Smithsonian Institution, a testament to the downfall of a species.

A Festival That's for the Birds

The residents of Hinckley don't know why the buzzards come, but they embrace their arrival every year with a festival.

Every March 15, the citizens of Hinckley Township celebrate the seasonal return of the buzzards with a huge festival. It's just like the return of the swallows to Capistrano in San Juan, California—only with more carrion. The event draws buzzard-loving crowds from around the world and includes events such as Buzzard Bingo, live music, crafts, and storytelling.

The buzzards—turkey vultures, to be specific—make their annual reappearance at the Buzzard Roost in Hinckley Reservation.

An officially appointed buzzard spotter scans the skies and clocks the arrival of the first bird. A local radio station sometimes hosts a lottery, and the contestant with the closest time to the first buzzard's arrival wins a prize. Once the first one shows up, the fun starts.

Why Hinckley?

No one seems to know why the buzzards show up, though there are several theories. One story suggests that the birds were drawn to the region following the big Hinckley Hunt of 1818, in which a group of men and boys went hunting for food and vermin and returned with more dead critters than they knew what to do with. It was a big mess for the townsfolk, and a big buffet for the buzzards.

Another theory is that the buzzards were drawn to the region in 1808 when the local Native Americans hanged a woman suspected of being a witch. Two years later, trapper William Coggswell reported seeing vultures flying over where the gallows stood. Whatever the case may be, buzzards have been good to Hinckley Township.

Beast of Burden

The amazingly adaptable camel can plod through the desert for a week without fluids—but don't attribute this water-conservation ability to that big hump.

With an unwieldy body that defies its life's mission and a disposition that often has it spitting at its owner, a camel's value as a "desert horse" seems questionable. Then there's that oversized hump, or humps in the case of the Bactrian camel. Hideous so far as aesthetics go, this natural canteen is the camel's true claim to fame. Because of it, the ungainly beast can travel with impunity in temperatures hot enough to fry an egg or kill a person. Or so many people believe.

In truth, this assertion is all wet. A camel does not store water in its hump. That bulge is composed primarily of fatty tissue that, when metabolized, serves as a source of energy. When this energy supply runs low because of a lack of nourishment, the hump shrinks considerably, sometimes to the point of flopping over to one side. On a healthy camel, however, the hump can weigh as much as 80 pounds.

A camel has a unique way of carrying and storing water—through its bloodstream. For this reason, it can go as many as eight days without a drink and can lose as much as 40 percent of its body weight before it feels ill effects. The amount it drinks when water is available—as much as 21 gallons in about ten minutes—would cause severe problems in most animals. What's more, a camel isn't too particular about the water it drinks. A muddy puddle that another animal might wrinkle its nose at would be slurped dry by a thirsty camel.

How Well Do You Understand Sharks?

They look frightening and strike fear into the hearts of nearly everyone who dips their toes into the ocean. Unfortunately, sharks are still one of the most misunderstood creatures on Earth. Do you know which rumors are true and which aren't?

Sharks are vicious man-eaters. False! People are not even on their preferred-food list. Every hunt poses a risk of injury to sharks, so they need to make every meal count. That's why they go for animals with a lot of high-calorie fat and blubber—

they get more energy for less effort. Humans are usually too lean and bony to be worth the risk.

Sharks are loners. That depends on the shark. Some species, such as the great white, are only rarely seen in the company of other sharks. However, many other species aren't so antisocial. Blacktip reef sharks hunt in packs, working together to drive  fish out of coral beds so every shark in the group gets a meal. Near Cocos Island off Costa Rica, hammerheads have been filmed cruising around in schools that consist of hundreds of sharks.

You have a greater chance of being killed by a falling coconut than by a shark. True! Falling coconuts kill close to 150 people every year. In comparison, sharks kill only five people per year on average. The International Shark Attack File estimates that the odds of a person being killed by a shark are approximately 1 in 264 million.

Sharks have poor vision, and most attacks are cases of mistaken identity. As popular as this belief is, it's wrong. Scientists have observed that sharks' behavior when they are hunting differs significantly from what most people report when a bite occurs. Sharks are extremely curious creatures and, since they don't have hands, they frequently explore their environment with the only things they have available—their mouths. Unfortunately for humans, a curious shark can do a lot of damage with a "test" nibble, especially if it's a big shark.

Most shark attacks are not fatal. True! There are about 60 shark attacks around the world each year, and, on average, just 1 percent of those are fatal. What we usually call "attacks" are only bites. Scientists report that an inquisitive shark that bites a surfboard (or an unlucky swimmer) shows far less aggression than when it is on the hunt and attacks fiercely and repeatedly.

Most shark attacks occur in water less than six feet deep. True! And the reason is obvious—that's where the majority of people are. It makes sense that most of the interactions between humans and sharks happen where the concentration of people is at its greatest.

Shark cartilage is an effective treatment for cancer. False! Anyone toting the benefits of shark cartilage as a nutritional supplement to cure cancer is selling snake oil. Multiple studies by Johns Hopkins University and other institutions have shown that shark cartilage has no benefit. This myth got started with the popular but incorrect notion that sharks don't get cancer. Sharks have to swim constantly or they drown. There are a few species that need to keep moving, but most sharks can still get oxygen when they're "motionless." They just open their mouths to draw water in and over their gills.

Most sharks present no threat to humans. True! There are more than 400 species of sharks, and approximately 80 percent of them are completely harmless to people. In fact, only four species are responsible for nearly 85 percent of unprovoked attacks: bull sharks, great white sharks, tiger sharks, and great hammerhead sharks.

The Impossible 'Possum

Let's face it: Opossums are weird. But here are some interesting facts that might change your mind about the unique opossum.

• Opossums are the only marsupial (i.e., a mammal that carries its young in a pouch) found in North America.

• The word "opossum" comes from the Algonquin word *apasum*, meaning "white animal." Captain John Smith used "opossum" around 1612, when he described it as a cross between a pig, a cat, and a rat.

• Although it is often colloquially called the "'possum," the opossum is completely different from a possum, an Australian marsupial.

• The opossum's nickname is "the living fossil," as it dates back to the dinosaurs and the Cretaceous Period, 70 million years ago. It is the oldest surviving mammal family on Earth.

• When cornered, opossums vocalize ferociously and show all 50 of their teeth, which they have more of than any other mammal. But they are lovers, not fighters, and prefer to run from danger.

• If trapped, they will "play 'possum," an involuntary response in which their bodies go rigid, and they fall to the ground in a state of shock. Their breathing slows, they drool a bit, and they release smelly green liquid from their anal sacs. This is enough to convince most predators the opossum is already dead, leaving it alone.

• Despite their predilection for eating anything—including rotting flesh—opossums are fastidious about hygiene. They bathe themselves frequently, including several times during each meal.

• Opossums are extremely resistant to most forms of disease and toxins, including rabies and snake venom, the latter probably due to their low metabolism.

• The idea that opossums mate through the female's nostrils is a myth. Although the male opossum has a forked (bifid) penis, he mates with the female in the normal manner. Conveniently, she has two uteri, so he deposits sperm into both of them at once.

Positively Platypus

What would you get if you could cross a duck with a beaver, a venomous snake, a chicken, and an otter? Well, it might look something like the duck-billed platypus.

What on Earth...?

The oldest platypus fossil (of the platypus in its current form) dates back more than 100,000 years. Indigenous to the south-eastern coast of Australia and Tasmania, the platypus is a monotreme, or egg-laying mammal. In fact, it is one of only two mammals that reproduce by laying eggs (the other is the echidna). In 1798, Captain John Hunter sent sketches and the skin of a platypus back to England that described a small, egg-laying animal with a pelt and a flat bill. They thought he was joking. British scientist Dr. George Shaw, thinking it was a hoax, tried to cut the pelt with scissors, expecting to find stitches attaching the bill to the body.

It's no wonder: The platypus is one strange-looking animal. With its fur and body shape, it resembles an otter. The tail looks like that of a beaver, but while beavers use their broad tails for propulsion, the platypus uses its tail for fat storage. It has four webbed feet, but only the front two are used for swimming. The rear feet are used for steering in water and aid in walking when on dry land.

Then there's the "duck bill." The platypus's rubbery snout is actually decidedly different than a duck's. The bill of a duck is hinged. The platypus's bill is a single piece of leathery skin with two nostrils on the top and a small mouth on the bottom.

High Tech Meets Monotreme

Platypuses have a unique sense of perception called electro-reception. Located inside the platypus's bill, electroreceptors can detect electrical fields that are generated by muscular contractions of other animals in the vicinity. Since the platypus swims with its eyes closed, the electroreceptors enable it to detect even the smallest movement of its prey. If it senses an oncoming attack from a predator, the male platypus can sting with an ankle spur that is loaded with poisonous venom. While it is not lethal to humans, the venom is powerful enough to kill dogs and other animals.

Dining and Breeding

The platypus is a semiaquatic mammal and spends up to 12 hours a day searching for food under water. In order to survive, it must eat up to 25 percent of its body weight every day. And because it thrives on insects, worms, larvae, freshwater shrimp, and other small organisms, the platypus has to be an accomplished diver and stay submerged for up to 40 seconds at a time. A bottom feeder, the platypus normally forages shallow river bottoms, rooting around for things to eat. While under-water, the platypus fills its cheeks with prey until it reaches the surface, where it takes time to enjoy its meal. At the end of the day, the platypus retires to a nest dug out of a riverbank and furnished with leaves, grass, and twigs. While the platypus is admittedly more awkward out of water, it's still agile enough to run along the ground.

When it comes to the birds and the bees, the platypus is definitely one of the more interesting critters out there. Many are confounded by the fact that the platypus lays eggs, but yet is not a bird. The female usually lays two eggs two weeks after mating, and the eggs are incubated for 10 to 14 days, as the mother holds the eggs to her body. Even more baffling, instead of feeding from its mother's teats like other mammals, the baby platypus gets its milk by lapping it up from the hairs and grooves in the mother's skin.

You'd be hard pressed to think of an animal that has more unusual characteristics than the duck-billed platypus. But hey, after 100,000 years spent plodding around on the planet, it must be doing something right!

Do Birds Get Tired in Flight?

Flying uses up a lot of energy. How do birds keep on truckin'?

Flight—especially migration—can be an exhausting experience for any bird. Reducing the amount of energy that is spent in the air is the primary purpose of a bird's body structure and flight patterns. Even so, migrating thousands of miles twice a year takes its toll on a bird's body, causing some to lose up to 25 percent of their body weight. How do they do it?

Large birds cut energy costs by soaring on thermal air currents that serve to both propel them and keep them aloft, which minimizes the number of times the beasts have to flap their wings. The concept is similar to a moving walkway at an airport: The movement of the current aids birds in making a long voyage faster while expending less energy.

Smaller birds lack the wingspans to take advantage of these currents, but there are other ways for them to avoid fatigue. The thrush, for instance, has thin, pointed wings that are designed to take it great distances while cutting down on the energy expended by flapping. Such small birds also have light, hollow bone structures that keep their body weights low.

If you've ever seen a gaggle of migrating geese, you likely noticed the distinctive V-formation that they take in flight.

They do this to save energy. The foremost goose takes the brunt of the wind resistance, while the geese behind it in the lines travel in the comparatively calm air of the leader's wake. Over the course of a migration, these birds rotate in and out of the leader position, thereby dispersing the stress and exhaustion.

While large birds routinely migrate across oceans, smaller birds tend to keep their flight paths over land—they avoid large bodies of water, mountain ranges, and deserts. This enables them to make the occasional pit stop.

Perhaps the most amazing avian adaptation is the ability to take short in-flight naps. A bird accomplishes this by means of unilateral eye closure, which allows it to rest half of its brain while the other half remains conscious. In 2006, a study of Swainson's Thrush—a species native to Canada and some parts of the United States—showed that the birds took hundreds of in-flight naps a day. Each snooze lasted no more than a few seconds, but in total, they provided the necessary rest.

How sweet is that? After all, who among us wouldn't want to take naps at work while still appearing productive?

A Lizard for the Ages

Horned toads may be the official state reptile of Texas, but one has gone well beyond its cold-blooded brethren to secure its own spot in Texas history.

In 1897, the cornerstone of the County Courthouse in Eastland was being placed. Among the gathered crowd were County Clerk Ernest Wood and his son, Will, who had a fondness for capturing lizards.

Horned toads, by the way, are lizards, not frogs, and have a rounded body and protrusions around their heads. Some types of horned toads actually squirt blood from the corners of their eyes, something that discourages predators but thrills many young boys.

During the placement of the cornerstone, several items were placed in a hollow in the marble cornerstone. County Clerk Wood got a wild hair and added Will's horned toad.

Is It a Miracle?

Thirty-one years later, more than 3,000 people gathered to see the reopening of that cornerstone to check on the lizard. When the old cornerstone was opened and the dusty and apparently lifeless horned toad was held aloft, it began kicking a leg and looking for breakfast.

Dubbed Old Rip, the horned toad lived another year during which it toured the United States and received a formal audience with President Calvin Coolidge. It was also featured in *Ripley's Believe It or Not*, and newsreels showed Rip's warty face across the land. Unfortunately, celebrity status proved to be too much. Old Rip returned home and died of pneumonia.

But it was carefully preserved in a tiny velvet-lined coffin in a glass case, becoming a miniature tourist attraction.

Even in death, Old Rip continued its adventures. At one point Rip was kidnapped, and at another, its leg fell off due to rough handling by a visiting politician. Old Rip may have been the inspiration for the Warner Brothers cartoon character "Michigan J. Frog," the tap dancing, singing frog that comes to life when a building is razed.

Nature's Nerds: The World's Brainiest Animals

It's notoriously difficult to gauge intelligence, both in humans and animals. Comparing animal IQs is especially tricky, since different species may be wired in completely different ways. But when you look broadly at problem-solving and learning ability, several animal brainiacs do stand out from the crowd.

Great Apes—Scientists generally agree that after humans, the smartest animals are our closest relatives: chimpanzees, gorillas, orangutans, and bonobos (close cousins to the common chimpanzee). All of the great apes can solve puzzles, communicate using sign language and keyboards, and use tools. Chimpanzees even make their own sharpened spears for hunting bush babies, and orangutans can craft hats and roofs out of leaves. One bonobo named Kanzi has developed the language skills of a three-year-old child—and with very little training. Using a computer system, Kanzi can "speak" around 250 words and can understand 3,000 more.

Dolphins and Whales—Dolphins are right up there with apes on the intelligence scale. They come up with clever solutions to complex problems, follow detailed instructions, and learn new information quickly—even by watching television. They also seem to talk to each other, though we don't understand their language. Scientists believe some species use individual "names"—a unique whistle to represent an individual—and that they even refer to other dolphins in "conversation" with each other. Researchers have also observed dolphins using tools. Bottlenose dolphins off the coast of Australia will slip their snouts into sponges to protect themselves from stinging animals and abrasion while foraging for food on the ocean floor. Marine biologists believe whales exhibit similar intelligence levels as well as rich emotional lives.

Elephants—In addition to their famous long memories, elephants appear to establish deep relationships, form detailed mental maps of where their herd members are, and communicate extensively over long distances through low-frequency noises. They also make simple tools, fashioning fans from branches to shoo away flies. Researchers have observed that elephants in a Kenyan national park can even distinguish between local tribes based on smell and clothing. The elephants are fine with one tribe but wary of  the other, and for good reason: That tribe sometimes spears elephants.

Parrots—People see intelligence in parrots more readily than in other smart animals because they have the ability to speak human words. But in addition to their famed verbal abilities, the birds really do seem to have significant brain power. The most famous brainy bird, an African gray parrot named Alex, who died in 2007, exhibited many of the intellectual capabilities of a five-year-old. He had only a 150-word vocabulary, but he knew basic addition, subtraction, spelling, and colors, and had mastered such concepts as "same," "different," and "none."

Monkeys—They're not as smart as apes, but monkeys are no intellectual slouches. For example, macaque monkeys can understand basic math and will come up with specific cooing noises to refer to individual objects. Scientists have also trained them to learn new skills by imitating human actions, including using tools to accomplish specific tasks. They have a knack for politics, too, expertly establishing and navigating complex monkey societies.

Dogs—If you're looking for animal brilliance, you might find it right next to you on the couch. Dogs are good at learning tricks, and they also demonstrate incredible problem-solving abilities, an understanding of basic arithmetic, and mastery of navigating complex social relationships. A 2009 study found that the average dog can learn 165 words, which is on par with a two-year-old child.

Freaky Friends

Some people are drawn to cute and cuddly puppies or furry little kittens while others like the idea of owning a unique one-of-a-kind pet. It's perfectly legal to keep some very unusual animals as pets (double-check before you buy one). But would you want to?

Serval—If you're a cat-lover, this might be the pet for you. Servals are larger than the average house cat, measuring three feet from head to tail and weighing anywhere between 15 and 40 pounds. They're regal animals but beware—they don't take to litter box training as well as most house cats.

Stick—No, we're not promoting a new Pet Rock fad here. A stick is a three- to four-inch-long insect that resembles, well, a stick. With proper care they can live for several years. Like the popular hermit crabs, their needs are simple and they're inexpensive to own. Sticks like company, so you if you've got the space, you might want to buy two.

Chinchilla—Chinchillas look like a cross between a rabbit and a squirrel and are known for their thick luxurious gray fur. Their coats are so thick and water-resistant that they clean themselves by rolling in dirt. While it may be legal to own a chinchilla, it is not legal to kill it to make its coat your own.

Wallaroo—Even the name is cute. If you want a pet the size of a 5th grader this may be the one for you. A cross between a kangaroo and a wallaby (in size and appearance) this Australian marsupial's typically lifespan is 15 to 20 years.

Kinkajou—At first glance, you might think this animal was a primate, but it's really a part of the raccoon family. Weighing in at a mere three to seven pounds, kinkajous love to hang by their tails. They have long and skinny tongues—which help them dine on honey and termites in the wild. And for a good parlor trick, kinkajous take the "moon walk" dance to a new level—they can turn their feet backwards to run in either direction.

Capybara—One look at this adorable animal (in a photo) and you'd probably be hooked. But would you feel differently if we told you the capybara is the world's largest rodent at four feet tall and more than 100 pounds? Think twice before you get one, though—they prefer a home with a swimming pool, can be territorial and have very large teeth. And did we mention that it's a rodent?

CHAPTER 3

THE ROOT OF ALL EVIL

Mansa Musa's Reverse Gold Rush

History's epic gold rushes were generally characterized by masses of people trekking to the gold. But in 1324, the legendary Mansa Musa bucked the trend by trekking masses of gold to the people—which had the result of severely depressing the Egyptian gold market.

If you could talk to a gold trader from fourteenth century Cairo, he might say that the worst time of his life occurred the day Mansa Musa came to town.

Musa, king of the powerful Mali Empire, stopped over in Cairo during his pilgrimage to Mecca. Arriving in the Egyptian metropolis in 1324, Musa and his entourage of about 60,000 hangers-on were anything but inconspicuous. Even more conspicuous was the 4,000-pound hoard of gold that Musa hauled with him.

While in Cairo, Musa embarked on a spending and gift-giving spree unseen since the pharaohs. By the time he was finished, Musa had distributed so much gold around Cairo that its value plummeted in Egypt. It would be more than a decade before the price of gold recovered from the Mali king's extravagance.

A Fool and His Money?—Not Musa

Musa's story conjures the old adage about fools and their money soon parting—especially when you consider he had to borrow money for the trip home. But Musa was no fool.

Musa ruled Mali from 1312 until 1337, and ushered in the empire's golden age. He extended Mali's power across sub-Saharan Africa from the Atlantic coast to western Sudan. Mali gained tremendous wealth by controlling the trans-Sahara trade routes, which passed through Timbuktu and made the ancient city the nexus of northwest African commerce. During Musa's reign, Mali exploited the Taghaza salt deposits to the north and the rich Wangara gold mines to the south, producing half the world's gold.

Musa's crowning achievement was the transformation of Timbuktu into one of Islam's great centers of culture and education. A patron of the arts and learning, Musa brought Arab scholars from Mecca to help build libraries, mosques, and universities throughout Mali. Timbuktu became a gathering place for Muslim writers, artists, and scholars from Africa and the Middle East. The great Sankore mosque and university built by Musa remain the city's focal point today.

Musa's Hajj Puts Mali on the Map

Musa's story is seldom told without mention of his legendary pilgrimage, or hajj, to Mecca.

The hajj is an obligation every Muslim is required to undertake at least once in their life. For the devout Musa, his hajj would be more than just a fulfillment of that obligation. It would also be a great coming-out party for the Mali king.

Accompanying Musa was a flamboyant caravan of courtiers and subjects dressed in fine Persian silk, including 12,000 personal servants. And then there was all that gold. A train of 80 camels carried 300 pounds of gold each. Five hundred servants carried solid-gold staffs.

Along the way, Musa handed out golden alms to the needy in deference to one of the pillars of Islam. Wherever the caravan halted on a Friday, Musa left gold to pay for the construction of a mosque. And don't forget his Cairo stopover. By the time he left Mecca, the gold was all gone.

But one doesn't dish out two tons of gold without being noticed. Word of Musa's wealth and generosity spread like wildfire. He became a revered figure in the Muslim world and inspired Europeans to seek golden kingdoms on the Dark Continent.

Musa's journey put the Mali Empire on the map—literally. European cartographers began placing it on maps in 1339. A 1375 map pinpointed Mali with a depiction of a black African king wearing a gold crown and holding a golden scepter in his left hand and a large gold nugget aloft in his right.

Running the Blockade

When President Lincoln set out to put an end to Southern trade, blockade running kept the Confederate economy alive.

After the attack on Fort Sumter, President Lincoln moved quickly to choke off Southern trade in the Atlantic and the

Gulf of Mexico. He believed that if he could blockade Confederate ports and shut down the Southern economy, he could put a speedy end to the rebellion. The catch was that this plan required a navy, and the Union had only a few ships at its disposal: just 27 steam ships and 44 sailing vessels to guard 3,350 miles of Confederate coastline.

While the North borrowed, built, or bought any ship it could, the South, realizing trade meant survival, did what it could to slip through the noose. It wasn't long before mariners stepped forward to take up the challenge of sneaking in and out of ports such as Galveston, Texas; Mobile, Alabama; and Wilmington, North Carolina.

The Business of Running

While a few blockade runners were commissioned by the Confederate government, most of the ships were owned by their captains or by syndicates of rich merchants. The most successful captains were often former officers of the U.S. Navy.

Because the South didn't have a manufacturing base, many of the runners were built in England. Their specifications combined the perfect mix of characteristics for a fast ship, sacrificing seaworthiness for speed and mobility. The *Banshee* was a prime example of this: A length of 214 feet and a width of only 20 feet allowed the ship to slice through the water. Burning hard Welsh anthracite coal so it wouldn't make smoke (most Southern mines brought up soft bituminous coal that left a telltale cloud behind the ship), the *Banshee* could move at speeds as fast as 18 knots, or 20.7 miles per hour. The ship had telescoping stacks to lower its silhouette and was painted a dull gray, one of the first instances of naval camouflage.

Cotton Currency

The Southern economy relied on supplying Europe with cotton. Blockade runners would be piled high with baled cotton for a dash to Bermuda or the Bahamas. There the cargo would be traded for Southern necessities, and the runner would speed back to any open Confederate port on the railroad, where those goods would be sold for a huge profit. As long as the system worked, everyone got rich. A captain could make $5,000 for one voyage, and even common sailors were paid $250, a fortune in the 1860s. But due to inflation, prices for goods shot up higher and higher as the blockade tightened. By 1864, it cost $300 Confederate for a barrel of flour and $40 Confederate for a pound of coffee.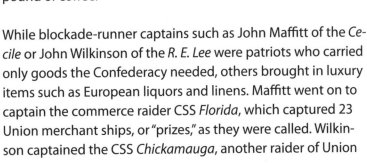

While blockade-runner captains such as John Maffitt of the *Cecile* or John Wilkinson of the *R. E. Lee* were patriots who carried only goods the Confederacy needed, others brought in luxury items such as European liquors and linens. Maffitt went on to captain the commerce raider CSS *Florida*, which captured 23 Union merchant ships, or "prizes," as they were called. Wilkinson captained the CSS *Chickamauga*, another raider of Union commerce.

Northern Incentives

Union sailors on the blockade line made only $16 per month, but they could also receive prize money for captured blockade runners and their cargoes, which were auctioned off at prize courts. As the Union navy grew from 264 ships by the end of 1861 to 588 ships by the end of 1863, more and more Southern runners were being captured. Not only were there more Union ships, but they were faster, with more experienced crews. As the Union armies worked their way through the South, fewer and fewer ports were left open to the Confederacy.

By early 1865, the Union took control of the last Confederate port, the one where it had all started—Charleston, South Carolina—and the days of blockade running were over. The *Banshee* was already long gone, captured on her ninth trip in 1863 and added to the Union blockade fleet.

Worth a Fortune: Very Rare U.S. Coins

Why are certain coins so valuable? Some simply have very low mintages, and some are error coins. In some cases (with gold, in particular), most of the pieces were confiscated and melted. Better condition always adds value.

The current mints and marks are Philadelphia (P, or no mark), Denver (D), and San Francisco (S). Mints in Carson City, Nevada (CC); Dahlonega, Georgia (D); and New Orleans (O) shut down long ago, which adds appeal to their surviving coinage. Here are the most prized and interesting U.S. coins.

1787 Brasher gold doubloon: It was privately minted by goldsmith Ephraim Brasher before the U.S. Mint's founding in 1793. The coin was slightly lighter than a $5 gold piece, and at one point in the 1970s it was the most expensive U.S. coin ever sold.

1792 half-disme (5¢ piece): Disme was the old terminology for "dime," so half a disme was five cents. George Washington supposedly provided the silver for this mintage. Was Martha the model for Liberty's image? If so, her hairdo suggests she'd been helping Ben Franklin with electricity experiments.

1804 silver dollar: Though actually minted in 1834 and later, the official mint delivery figure of 19,570 refers to the 1804 issue. Watch out—counterfeits abound.

1849 Coronet $20 gold piece: How do you assess a unique coin's value? The Smithsonian owns the only authenticated example, the very first gold "double eagle." Why mint only one? It was a trial strike of the new series. Rumors persist of a second trial strike that ended up in private hands; if true, it hasn't surfaced in more than 150 years.

1870-S $3 gold piece: Apparently, only one (currently in private hands) was struck, though there are tales of a second one placed in the cornerstone of the then-new San Francisco Mint building (now being renovated as a museum). If the building is ever demolished, don't expect to see it imploded.

1876-CC 20-cent piece: Remember when everyone confused the new Susan B. Anthony dollars with quarters? That's what comes of ignoring history. A century before, this 20-cent coin's resemblance to the quarter caused similar frustration.

1894-S Barber dime: The Barber designs tended to wear quickly, so any Barber coin in great condition is scarce enough. According to his daughter Hallie, San Francisco Mint director John Daggett struck two dozen 1894-S coins, mostly as gifts for his rich banker pals. history.

1907 MCMVII St. Gaudens $20 gold piece: This is often considered the loveliest U.S. coin series ever. Its debut featured the year in Roman numerals, unique in U.S. coinage. The first, ultra-high-relief version was stunning in its clarity and beauty, but it proved too time-consuming to mint, so a less striking (but still impressive) version became the standard.

1909-S VDB Lincoln cent: It's a collectors' favorite, though not vanishingly rare. Only about a fourth of Lincoln pennies from the series' kickoff year featured designer Victor D. Brenner's initials on the reverse; even now, an occasional "SVDB" will show up in change.

1913 Liberty Head nickel: This coin wasn't supposed to be minted. The Mint manufactured the dies as a contingency before the Buffalo design was selected for 1913. Apparently, Mint employee Samuel W. Brown may have known that the Liberty dies were slated for destruction and therefore minted five of these for his personal gain.

1913-S Barber quarter: Forty thousand of these were made—the lowest regular-issue mintage of the twentieth century. Some Barbers wore so flat that the head on the obverse was reduced to a simple outline.

1915 Panama-Pacific $50 gold piece: This large commemorative piece was offered in both octagonal and round designs.

1916 Liberty Standing quarter: This coin depicts a wardrobe malfunction . . . except by design! Many were shocked when the new coin displayed Lady Liberty's bared breast. By mid-1917, she was donning chain mail. Like the Barber quarter before it, the Liberty Standing wore out rapidly. With only

52,000 minted in 1916, the series' inaugural year, a nice specimen will set you back nearly $40,000.

1933 St. Gaudens $20 gold piece: This coin is an outlaw. All
of the Saint's final mintage were to be melted down—and
most were. Only one specific example is legal to own; other
surviving 1933 Saints remain hidden from the threat of Treasury confiscation. The legal one sold in 2002 for an incredible
$7.6 million.

1937-D "three-legged" Buffalo nickel: A new employee at
the Denver Mint tried polishing some damage off a die with
an emery stick. He accidentally ground the bison's foreleg off,
leaving a disembodied hoof. No telling exactly how many were
struck, but they sure look funny.

The Downside to Collecting Coins

Faced with a frustrating coin shortage, a New York barkeep
took matters into his own hands. By the second year of the
Civil War, Northern merchants were finding it more and more
difficult to make change for their customers. Americans, set on
edge by the war and worried that financial catastrophe was
just around the corner, started to hang onto their coins. At that
time, coins were cast from actual gold and silver, commodities
that would remain valuable even if the government that had
stamped its name on the coins were to crumble. This paranoia
led to a coin shortage, which made everyday life difficult. The
machinery of the American economy threatened to grind to
a halt.

Lindenmueller Coins

In 1863, frustrated with the lack of coins, New York bartender Gustavus Lindenmueller took matters into his own hands.

In order to make change for his patrons, he literally made change—about one million cents' worth—and began to distribute it. The coins were simple, featuring his bearded face on the front and a frothy mug of ale on the back. The large volume of coins Lindenmueller dumped into circulation in the city made them popular with a public equally annoyed by the lack of coins. Many businesses accepted them, despite the facts that they had absolutely no true value and that there was no indication they would be honored if ever put to the test. Streetcars began to accept them, for instance, because exact change made traveling easier than using cumbersome paper money.

Filling a Void

Lindenmueller coins were just one of several tokens privately created in the war's early years. It's believed nearly 25 million such tokens were minted across the North. They filled a void that the government had not been able to account for and reflected American ingenuity and patriotism. Many were adorned with spirited slogans and pro-Union sentiments. It may seem strange that so many people and even companies accepted these dubious tokens and private currencies, but in the midst of the Civil War, they had little choice. As long as everyone agreed to treat the tokens as though they had real worth, the needs of daily life could be met.

THE ROOT OF ALL EVIL

Put to the Test

The situation could not remain like this forever. The Third Avenue Railroad served the city of New York with streetcar service, and rather than go bankrupt due to the lack of coins for train fare, they decided to play along with the Lindenmueller tokens, accepting them from riders in lieu of actual U.S. currency. Lindenmueller had given them out as change instead of real U.S. coins, so the railroad apparently assumed that, as a reputable and honorable tavern keeper, he would honor them himself. When the company presented him with a large number of his tokens in an attempt to redeem them for actual money, however, Lindenmueller laughed in their faces.

Although he was happy to give them out, he had no intention of accepting them himself—at least not in bulk. Because the railroad had participated in the scheme willingly, and because Lindenmueller had never made explicit promises to honor the coins, there was absolutely nothing the railroad could do. Problems such as this finally stirred the government to act against these illegal currencies.

Two-Cents Worth

The U.S. Congress undertook a two-pronged attack on the Civil War tokens, rendering them both unnecessary and illegal. Anyone caught using or creating such currency would be subject to the same treatment as counterfeiters or forgers—a significant fine and possibly five years in jail. Congress also decided to beat those tokenmakers at their own game by introducing the two-cent coin. In many ways, it was modeled after the very tokens it was created to eliminate.

Deadly Bling?: The Curse of the Hope Diamond

Diamonds are a girl's best friend, a jeweler's meal ticket, and serious status symbols for those who can afford them. But there's one famous diamond whose brilliant color comes with a cloudy history. The Hope Diamond is one of the world's most beautiful gemstones—and one that some say causes death and suffering to those who possess it. So is the Hope Diamond really cursed? Evidence says "no," but there have been strange coincidences.

It's believed that this shockingly large, blue-hued diamond came from India several centuries ago. At this time, the exceptional diamond was slightly more than 112 carats, which is enormous. (On average, a diamond in an engagement ring ranges from a quarter to a full carat.) According to legend, a thief stole the diamond from the eye of a Hindu statue, but scholars don't think the shape would have been right to sit in the face of a statue. Nevertheless, the story states that the young thief was torn apart by wild dogs soon after he sold the diamond, making this the first life claimed by the jewel.

Courts, Carats, and Carnage

In the mid-1600s, a French jeweler, Tavernier, purchased the diamond in India and kept it for several years without incident before selling it to King Louis XIV in 1668, along with several other jewels. The king recut the diamond in 1673, taking it down to 67 carats. This new cut emphasized the jewel's clarity, and Louis liked to wear the "Blue Diamond of the Crown" around his neck on special occasions. He, too, owned the gem-stone without much trouble.

More than a hundred years later, France's King Louis XVI possessed the stone. In 1791, when the royal family tried to flee the country, the crown jewels were hidden for safekeeping, but they were stolen the following year. Some were eventually returned, but the blue diamond was not.

King Louis XVI and his wife Marie Antoinette died by guillotine in 1793. Those who believe in the curse are eager to include these two romantic figures in the list of cursed owners, but their deaths probably had more to do with the angry mobs of the French Revolution than a piece of jewelry.

Right This Way, Mr. Hope

It is unknown what happened to the big blue diamond from the time it was stolen in France until it appeared in England nearly 50 years later. When the diamond reappeared, it wasn't the same size as before—it was now only about 45 carats. Had it been cut again to disguise its identity? Or was this a new diamond altogether? Because the blue diamond was so unique in color and size, it was believed to be the diamond in question.

In the 1830s, wealthy banker Henry Philip Hope purchased the diamond, henceforth known as the Hope Diamond. When he died (of natural causes) in 1839, he bequeathed the gem to his oldest nephew, and it eventually ended up with the nephew's grandson, Francis Hope.

Francis Hope is the next person supposedly cursed by the diamond. Francis was a notorious gambler and was generally bad with money. Though he owned the diamond, he was not allowed to sell it without his family's permission, which he finally got in 1901 when he announced he was bankrupt.

It's doubtful that the diamond had anything to do with Francis's bad luck, though that's what some believers suggest.

Coming to America

Joseph Frankel and Sons of New York purchased the diamond from Francis, and by 1909, after a few trades between the world's most notable jewelers, the Hope Diamond found itself in the hands of famous French jeweler Pierre Cartier. That's where rumors of a curse may have actually originated. Allegedly, Cartier came up with the curse concept in order to sell the diamond to Evalyn Walsh McLean, a rich socialite who claimed that bad luck charms always turned into good luck charms in her hands. Cartier may have embellished the terrible things that had befallen previous owners of his special diamond so that McLean would purchase it—which she did. Cartier even inserted a clause in the sales contract, which stated that if any fatality occurred in the family within six months, the Hope Diamond could be exchanged for jewelry valued at the $180,000 McLean paid for the stone. Nevertheless, McLean wore the diamond on a chain around her neck constantly, and the spookiness surrounding the gem started picking up steam.

Whether or not anything can be blamed on the jewel, it certainly can't be denied that McLean had a pretty miserable life starting around the time she purchased the diamond. Her eldest son died at age nine in a fiery car crash. Years later, her 25-year-old daughter killed herself. Not long after that, her husband was declared insane and was committed to a mental institution for the rest of his life. With rumors swirling about the Hope Diamond's curse, everyone pointed to the necklace when these terrible events took place.

The Root of All Evil

In 1947, when McLean died (while wearing the diamond) at age 60, the Hope Diamond and most of her other treasures were sold to pay off debts. American jeweler Harry Winston forked over the $1 million asking price for McLean's entire jewelry collection.

Hope on Display

If Harry Winston was scared of the alleged curse, he didn't show it. Winston had long wanted to start a collection of gemstones to display for the general public, so in 1958, when the Smithsonian Institute started one in Washington, D.C., he sent the Hope Diamond to them as a centerpiece. These days, it's kept under glass as a central figure for the National Gem Collection at the National Museum of Natural History. So far, no one's dropped dead from checking it out.

Haunting Up Cash

Every October, thousands of haunted attractions spring up across North America. For the price of admission, you get thrilled, chilled, and utterly spooked—and the organizers rake in the profits.

Around 2,000 haunted attractions are produced in the United States every year. Ticket sales, vendors, construction costs, decorations, and other supplies make this approximately a $500 million industry, which isn't bad for a business model that offers its product for just one month a year. The scope of the haunted attraction industry has become extremely impressive as well: Haunted houses today look very different than they used to.

Many haunted attractions were originally staged for charitable causes. Decades ago, nonprofit groups often organized "haunted houses" to raise money for sick kids or needy folks in the community. (One Jaycees chapter in Durham, North Carolina, celebrated its 38th haunted house in 2010; in recent years, it has annually raised more than $10,000 for charity.)

Many groups that create haunted attractions still give at least a portion of their proceeds to charity, but over time, "scare houses" have become actual businesses run by companies that focus solely on entertainment value . . . and profits. To get people in the door, haunted houses have gotten scarier and more complex; staging them takes a lot of creativity, time, and money.

Freak-You-Out Economics

The initial investment needed to launch a successful haunted attraction can be around $250,000. This may sound like a lot, but the designers, carpenters, performers, safety teams, cleaning crews, and Web and marketing professionals needed to create a blockbuster haunted attraction all cost money. Many of the necessary props and costumes are specialty items that far exceed most people's budgets, which is why those who wish to run a successful haunted house typically pursue loans and capital investments.

The good news is that at around $15 (or more) per head, several hundred people lining up to move through your haunted house each night during the month of October makes it likely that you'll turn a hefty profit. To attract enough people, some proprietors enlist the help of metal bands and freak show-style performers on opening night. For example, Marilyn Manson was once booked by a fright house in New Orleans that wanted to bring in its biggest crowd ever.

Cashing In on the Ghosts

If you're interested in the business of scaring people silly, plenty of Web resources are available. You can also attend the Midwest Haunters Convention, which is held each June. Annually, around 2,000 people attend this event to gather industry information, stay abreast of the latest trends, and contact vendors that sell everything from fake bloody limbs to dead-bride costumes.

Consumers are increasingly critical of those who try to earn their spending money, so proprietors of haunted attractions have their work cut out for them. Competition can be fierce: Haunts such as the Haunted Hoochie in Pataskala, Ohio, spans some 40 acres, and in 2010, the Lewisburg Haunted Cave in Lewisburg, Ohio, set a Guinness World Record for "Longest Walk-Through Horror House," boasting 3,564 linear feet of fright.

These days, it's typical for haunted attractions to reach 40,000 square feet or more—Canton, Ohio's Factory of Terror spans some 55,000 square feet—so make sure that you've got plenty of space. Those who love the haunted attraction experience are willing to travel to find the cream of the crop, so the scarier and more realistic, the better.

Unless, of course, it's too scary: In 2000, a woman filed a lawsuit against Universal Studios in Orlando, Florida, claiming that its annual Halloween Horror Nights attraction was so terrifying that she suffered emotional damage as a result; the outcome of her suit is unknown.

9 Really Odd Things Insured by Lloyd's of London

1. In 1957, world-famous food critic Egon Ronay wrote and published the first edition of the *Egon Ronay Guide to British Eateries*. Because his endorsement could make or break a restaurant, Ronay insured his taste buds for $400,000.

2. In the 1940s, executives at 20th Century Fox had the legs of actress Betty Grable insured for $1 million each. After taking out the policies, Grable probably wished she had added a rider to protect her from injury while the insurance agents fought over who would inspect her when making a claim.

3. While playing on Australia's national cricket team from 1985 to 1994, Merv Hughes took out an estimated $370,000 policy on his trademark walrus mustache, which, combined with his 6'4" physique and outstanding playing ability, made him one of the most recognized cricketers in the world.

4. Representing the Cheerio Yo-Yo Company of Canada,13-year-old Harvey Lowe won the 1932 World Yo-Yo Championships in London and toured Europe from 1932 to 1935. He even taught Edward VIII, then Prince of Wales, how to yo-yo. Lowe was so valuable to Cheerio that the company insured his hands for $150,000!

5. From 1967 to 1992, British comedian and singer Ken Dodd was in the *Guinness World Records* for the world's longest joke-telling session—1,500 jokes in three and a half hours. Dodd has sold more than 100 million comedy records and is famous for his frizzy hair, ever-present feather duster, and extremely large buckteeth. His teeth are so important to his act that Dodd had them insured for$7.4 million, surely making his insurance agent grin.

6. During the height of his career, Michael Flatley—star of *Riverdance* and *Lord of the Dance*—insured his legs for an unbelievable $47 million. Before becoming the world's most famous Irish step dancer, the Chicago native trained as a boxer and won the Golden Gloves Championship in 1975, undoubtedly dazzling his opponents with some extremely fast footwork.

7. The famous comedy team of Bud Abbott and Lou Costello seemed to work extremely well together, especially in their famous "Who's on First?" routine. But to protect against a career-ending argument, they took out a $250,000 insurance policy over a five-year period. After more than 20 years together, the team split up in 1957—not due to a disagreement, but because the Internal Revenue Service got them for back taxes, which forced them to sell many of their assets, including the rights to their many films.

8. Rock and Roll Hall of Famer Bruce Springsteen is known to his fans as The Boss, but Springsteen knows that he could be demoted to part-time status with one case of laryngitis. That's why in the 1980s he insured his famous gravelly voice for $6 million. Rod Stewart has also insured his throat, and Bob Dylan has a similar policy to protect his vocal cords for that inevitable day when they stop blowin' in the wind.

9. Before rock 'n' roll, a popular type of music in England in the 1950s was skiffle, a type of folk music with a jazz and blues influence played on washboards, jugs, kazoos, and cigar-box fiddles. It was so big at the time that a washboard player named Chas McDevitt tried to protect his career by insuring his fingers for $9,300. It didn't do him much good because skiffle was replaced by rock 'n' roll, washboards by washing machines, and McDevitt by McCartney.

Lotto Trouble!

Everyone fantasizes about winning the lottery and living the good life. But for every lucky winner who achieves the dream, there's another whose life is turned upside down. These true-life lottery horror stories may make you think twice about buying that scratch-off ticket.

1. In 1988, William Post won $16.2 million in the Pennsylvania Lottery. Unfortunately, Post's good luck brought out the worst in his friends and family, including his brother, who tried to have Post killed for the inheritance. Post survived his brother's murderous intent only to be successfully sued by a former girlfriend for a share of his winnings. Post spent his money like a drunken sailor until he was $1 million in debt and forced to declare bankruptcy. When he died in 2006, Post had been living on his Social Security for several years.

THE ROOT OF ALL EVIL

2. Everyone believes that winning the lottery will bring an end to their problems, but for Billie Bob Harrell Jr., that wasn't the case. After scoring $31 million in the Texas Lottery, Harrell spent big on cars, real estate, and gifts for family and friends. But wealth apparently couldn't buy the happiness Harrell was seeking, so he took his own life just two years after cashing that winning ticket.

3. Evelyn Adams was a two-time winner, hitting it big in the New Jersey Lottery in 1985 and 1986 for a total of more than $5 million. But Adams couldn't control her gambling habit and quickly frittered her winnings away on the slots.(She also gave away large sums to family and friends.) By 2001, the former multimillionaire was living in a trailer.

4. Willie Hurt won $3.1 million in the Michigan Lottery in 1989. Two years later, he was penniless and facing a murder charge after squandering his fortune on a costly divorce and crack cocaine.

5. Jeffrey Dampier won a whopping $20 million in the Illinois Lottery and spent lavishly on friends and family—including his sister-in-law, Victoria Jackson. Hungry for a bigger chunk of Dampier's fortune, Jackson and her boyfriend lured Dampier to Jackson's apartment, then kidnapped and murdered him. Jackson was sentenced to life in prison.

6. In December 2002, West Virginian Jack Whittaker won $314 million in the largest undivided Powerball jackpot in lottery history. He had nothing but good intentions for his money and gave generously to his church, his friends, and various civic organizations. But trouble seemed to follow Whittaker after he won the big one. Strangers hounded him for money, and some even threatened his family when he refused. He was hit with a variety of lawsuits and once was robbed of $500,000 in cash. But the greatest tragedy was when Whittaker's teenage granddaughter died of a drug overdose, her habit funded by the generous allowance she received from her grandfather.

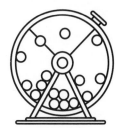

7. Wanda Rickerson won more than half a million dollars in the Georgia Lottery in 2003. Three years later, the former sheriff's office administrative clerk was ordered to pay $84,000 to Columbia County after pleading no contest to theft and insurance charges. Rickerson's crime? Embezzling $56,000 from an inmate trust account she supervised at the Columbia County Detention Center.8. An anonymous British man who won a sizable fortune in the national lottery was the victim of a home invasion following his windfall. Masked thugs wielding machetes burst into the man's home and held him, his wife, and his young son at knifepoint while they ransacked the place. The thieves escaped with jewelry and numerous personal items.

8. In 2001, Victoria Zell and her husband won an $11 million jackpot in the Minnesota Lottery. But Zell's good fortune didn't last long. In 2005, she was sent to prison for an alcohol-related car crash that killed one motorist and paralyzed another.

9. Michael Carroll's $17 million fortune couldn't keep him out of prison either. In 2006, the British lottery winner was sentenced to nine months in the hoosegow for going berserk in a disco.

THE ROOT OF ALL EVIL

The Whiskey Ring

In September 1869, James Fisk and Jay Gould started the Black Friday panic when they attempted to corner the gold market. The scandal tarnished President Ulysses Grant's reputation, and he started to lose support from some of his fellow Republicans in Missouri. So, Grant sent General John McDonald–the supervisor of the U.S. Treasury Department's internal revenue for the St. Louis area–to Missouri to round up some support for the struggling president. But instead of turning attention to something other than the controversy surrounding Grant's administration, McDonald introduced the country to a brand-new scandal.

Don't Drink and Bribe

In an effort to raise money for Grant, McDonald–along with what would grow to hundreds of other accomplices–partnered up with whiskey distillers in a scheme to benefit both parties. The distillers would pay federal agents huge bribes, and in return, the agents helped the distillers evade taxes. Tax on whiskey was 70 cents per gallon; but if a distiller bribed a federal agent, his whiskey was only taxed at 35 cents a gallon, and he could make impressive profits.

The Whiskey Ring, as it came to be known, spread from St. Louis to Chicago to New Orleans to Washington D.C. The plotters in the conspiracy included politicians, reporters, revenue service agents, shopkeepers, and, of course, whiskey distillers. Even Grant's private secretary and good friend, General Orville E. Babcock, participated in the ring. But with so many people in on the conspiracy, it was only a matter of time before the whole thing came crashing down.

The First Special Prosecutor

In 1875, word of the Whiskey Ring reached the U.S. Secretary of the Treasury, Benjamin Bristow. Bristow had already seen the effects of the Black Friday panic and the Credit Mobilier scandal, and he'd had enough of the corruption. Working secretly and without Grant's knowledge, Bristow used information he'd heard from members of the ring to conduct raids across the country on May 10, 1875. More than 200 men were indicted—with 110 eventual convictions, in-

cluding McDonald—and $3 million in taxes was recovered.After the ring was exposed, Grant appointed John B. Henderson as a special prosecutor—the country's first—to avoid any appearance of a conflict of interest. One of Henderson's indictments was Babcock, but Grant's longtime friend insisted to the president that he was innocent. When Henderson not only refused to accept Babcock's appeal of innocence but also suggested that Grant might have known about the ring, Grant fired him and replaced him with James Broadhead.

The Damage Is Done

Grant himself testified during Babcock's trial, and the president's secretary became the only major player in the Whiskey Ring to win an acquittal. However, the public's trust was gone, and Babcock was forced to resign his position. Bristow resigned as well, as members of Grant's cabinet—disgruntled that their friends and associates had been scrutinized—refused to speak to him. And Grant, although not involved in the Whiskey Ring himself, found his reputation marred by so many scandals

THE ROOT OF ALL EVIL

during his presidency. He left office in 1877 for less controversial pursuits: traveling around the world with his family.

A Charlatan of Epic Proportions

"Greed is good," said Gordon Gekko in Oliver Stone's 1987 hit movie *Wall Street*. Greed resides at the center of the financial industry. For the powerful real-life stockbroker Bernie Madoff, greed knew no boundaries.

Madoff, a well-respected broker who became chairman of the Nasdaq in 1990 and served in the position in 1991 and 1993, orchestrated the largest Ponzi scheme in history: an estimated $65 billion fraud. He conned thousands of investors and would later pled guilty to 11 felony charges—including money laundering, perjury, and fraud—and earn a prison sentence of 150 years.

Madoff was, according to the *New York Times*, a "charlatan of epic proportions, a greedy manipulator so hungry to accumulate wealth that he did not care whom he hurt to get what he wanted."

Small Beginnings

Madoff founded Bernard L. Madoff Securities in 1960. He started his firm with a paltry $5,000 he saved from lifeguarding. His wife's father allowed Madoff to work out of his Manhattan accounting firm. A market maker, Madoff dealt in over-the-counter penny stocks. He was also, Madoff would later recall in an interview, a "little Jewish guy from Brooklyn" who felt like he was on the outside looking in.

Madoff steadily grew his business—and reputation—as he embraced new trading technology and crafted friendships with industry regulators. While traders described him as obsessive, paranoid, secretive, and manipulative, Madoff earned the trust of employees, investors, and Wall Street. At the same time, Madoff was overseeing a massive con involving fraudulent transactions on an epic scale.

He later claimed that a handful of powerful clients known as the "Big Four" forced him into a Ponzi scheme in the early 1990s.

A Confession

In the late 1990s, Frank Casey, an investment firm executive, asked a colleague to look into Madoff's trades. The colleague, Harry Markopolos, quickly became suspicious and suspected a Ponzi scheme. Casey told PBS *Frontline* that Markopolos compared Madoff's returns to a baseball player "hitting .925 straight for 10 years in a row." Markopolos sent an eight-page memo to the SEC, but the agency did not follow up with an investigation. He wrote additional memos to the SEC, and in January 2006, the SEC launched an investigation.

Finally, at the height of the 2008 financial crisis, Madoff's jig was up.

On December 10, 2008, Madoff allegedly confessed to his sons that his business was a massive Ponzi scheme. The next day, authorities arrested Madoff on one count of securities fraud, and he was released on $10 million bail. In June 2009, a federal judge sentenced Madoff to 150 years in prison. He did not appeal the sentence.

THE ROOT OF ALL EVIL

Taking Responsibility?

The Department of Justice announced in December 2017 that it had begun to return money to Madoff's victims, including thousands of respected individuals and institutions. The initial distribution included $772.5 million, a fraction of the more than $4 billion in assets recovered for the victims. An additional $504 million was announced in April 2018.

In a 2011 interview with Barbara Walters, Madoff said he has no fear because "I'm no longer in control of my own life." He told Walters that he took full responsibility for his crimes, but he said, "Nobody put a gun to my head. I never planned to do anything wrong. Things just got out of hand."

Even prison hasn't keep Madoff's profit-motivating instincts at bay, however. At one point, the aging criminal reportedly purchased hot chocolate packets from the commissary and sold them for a profit in the prison yard.

Curious World Currencies

Paper, coins, and plastic are what we use as money today, but that wasn't always the case. Throughout history, people have used various animals, vegetables, and minerals to conduct business.

Cows represent the oldest of all forms of money, dating from as early as 9000 BC. The words "capital," "chattels," and "cattle" have a common root, and the word "pecuniary" (meaning "financial") comes from *pecus*, the Latin word for cattle. But cattle weren't the only livestock used as legal tender: Until well into the twentieth century, the Kirghiz (a Turkic ethnic group found primarily in Kyrgyzstan) used horses for large exchanges, sheep for lesser trades, and lambskins for barters that required only small change.

Cowry shells—marine snails found chiefly in tropical regions—were the medium of exchange used in China around 1200 BC. These shells were so widely traded that their pictograph became the symbol for money in the written language. The earliest metallic money in China were cowries made of bronze or copper.

Throughout history, salt and pepper have been used as money, both for their value as seasonings and preservatives and for their importance in religious ceremonies. In ancient Rome, salt was used as money, and the Latin word for salt (sal) is the root of the word "salary." Roman workers were paid with salt, hence the expression "worth one's salt." Pepper was also used as a form of payment. During the Middle Ages in England, rent could be paid in peppercorns, and the term "peppercorn rent" refers to the smallest acceptable payment.

The largest form of money is the 12-foot limestone coins from the Micronesian island of Yap. The value of the coin was determined by its size—the 12-foot rounds weighed several tons. Displaying a large coin outside your home was a sign of status and prestige.

Because of the coins' size and immobility, islanders would often trade only promises of ownership instead of actually exchanging them. Approximately 6,800 coins still exist around the island, though the U.S. dollar is now the official currency.

CHAPTER 4

CULTURAL CURIOSITIES

Ed Wood: Made in Hollywood, USA

Only in Hollywood, the land of make-believe, could Ed Wood realize his dream of making movies; only in Hollywood could his enthusiasm, drive, and perseverance overcome his lack of cinematic skill. His gift was not in making movies but in persuading people that he could make movies.

Edward D. Wood, Jr. wrote, directed, and produced some of the worst sci-fi and horror films of the 1950s—or any decade. Wood had the uncanny ability to convince people to sink money into his pictures, which often involved writing a new part in the film for the prospective investor.

Background

Born in Poughkeepsie, New York, in 1924, young Wood quickly found his passion. As a boy, he sat enthralled in darkened movie theaters from open to close. Wood's mother, who wanted a daughter rather than a son, often dressed Ed in girls' clothing, which was the genesis of his lifelong habit of cross-dressing. He joined the Marines at 17, later claiming to have often worn a bra and panties under his uniform during his World War II tour of duty. After the war, Wood settled in Hollywood to tilt at his own personal windmill.

Movies

Most notable of Wood's films was the 1953 autobiographical fantasy of cross-dressing, *Glen or Glenda?* Wood starred in it himself but used the pseudonym Daniel Davis.

His 1955 *Bride of the Monster* was a typical "mad scientist" flick, complete with laboratory (including, literally rather than proverbially, the kitchen sink) and a giant killer octopus. The octopus had been stolen from a prop warehouse—the only catch was that Wood forgot to pilfer the motor that moved the rubber tentacles, as well. *Night of the Ghouls*, shot in 1958, was a sequel of sorts to *Bride of the Monster*.

Wood's masterpiece, however, was the 1959 epic *Plan 9 from Outer Space*. Starting with a few minutes of silent footage, Wood made a very silly, yet very watchable, sci-fi film. Tagged by many as "the worst  movie ever made," the film's appeal stemmed from its intense sincerity. *Plan 9* never intentionally winked at the camera— never mind the cardboard tombstones that fell over, the shower curtain that passed for the cockpit of an airplane, or the tiny cemetery crypt that seemed to hold more people than a taxi full of clowns.

An Oddball Acting Troupe

Each of Wood's films featured a unique cast of characters, both on- and off-screen. John "Bunny" Breckenridge was a proud homosexual who dreamed of undergoing permanent gender reassignment. Blonde, beautiful Dolores Fuller was

Wood's girlfriend and confidant. White-haired eccentric The Amazing Criswell had startled TV audiences with his incredible (and usually inaccurate) predictions. Kenne Duncan was known in Hollywood as the "meanest man in the movies," and many believed him to be the most lecherous, as well. Buxom, wasp-waisted Maila Nurmi made her mark on LA television by hosting a horror movie show as the gaunt and dark-haired Vampira. Hulking Tor Johnson was a professional wrestler called the "Super Swedish Angel"—his nearly unintelligible Swedish accent (and total lack of acting talent) didn't seem to bother Wood at all.

Most amazing of Wood's troupe, however, was the aging, frail Bela Lugosi. A huge theater star in Europe and a hit on Broadway in the late 1920s in the stage play *Dracula*, Lugosi naturally went on to portray the undead count in Universal's 1931 film of the same name. It was a high point that he would never again reach. He worked steadily through the 1930s and '40s, appearing (but seldom starring) in roles as gangsters, doctors (mad and sane), butlers, and servants. He eventually became a comedic foil for the Bowery Boys in several films. By the 1950s, Lugosi was in his 70s, ravaged by an addiction to painkillers.

Wood greatly admired Lugosi and gave the elderly actor friendship and a sense of being wanted once again. Too old to play the romantic leads that he once coveted, Lugosi was cast instead as a godlike character in *Glen or Glenda*? Wood continued to keep his Hungarian friend active in other films and shot footage in 1956 for a new film called *The Vampire's Tomb* in which, once again, Lugosi would rise as a blood-thirsty vampire. But Bela passed away after shooting only a few minutes of footage, some of it in his revered Count Dracula cape.

Wood Had a Plan

While Wood lost a great friend in Lugosi, his resolve was firm, and he wrote a brand-new script to incorporate the footage that Lugosi had already shot. *Grave Robbers from Outer Space* would be for Ed Wood what *Citizen Kane* was for Orson Welles: a masterpiece. Wood shot additional scenes, using local chiropractor Dr. Tom Mason as Bela's double. The fact that Mason looked nothing like Lugosi seemed to make no difference to Wood. Money problems led Wood to affiliate himself with a local Baptist church, which agreed to finance the film. As usual, parts were written into the film to accommodate the desires of the parish's would-be thespians. The church elders found the title reference to grave robbing to be in poor taste, so the film was renamed *Plan 9 from Outer Space* (though Criswell introduced the film with the original title). With a small, unpromising premiere, Wood had accomplished what he set out to do—he finished the movie as a tribute to his fallen friend.

Legacy

Consider that some people like fast-food restaurants while others prefer haute cuisine. But as long as one doesn't expect four-star fare from a fast-food hamburger, the flavor can be very palatable. Wood's work was never very good cinema, but his films are entertaining. They seem to improve with age—not in quality but in watchability. The more viewers get to know this eccentric film writer, producer, director, and sometimes actor, the more they can accept his work for what it is—no more, no less. Wood's films were born not from talent but from a deep passion for the movies.

Lulu: Idol of Germany

From Hollywood "It Girl" to heyday has-been in less than five years—this was the fate cast by Hollywood studio bosses, who were unable to reel in the rebellious and outspoken Louise Brooks. But the "girl in the black helmet," whose trademark shiny black bob has since been copied by every woman and starlet alike, didn't prove so easy to throw aside.

Louise Brooks, born Mary Louise Brooks to an attorney father and artistically eccentric mother in Kansas in 1906, picked up her contentious temperament early. The philosophy, if that's not too formal a term for it, of the Brooks household was every child for herself. Mother Myra Brooks was known to sit back and laugh as her "squalling brats" took to fisticuffs to settle their disputes. Fighting, it seemed, was in Louise's blood. She started her serious performing career at age 15 as a dancer in the prestigious Denishawn troupe but was later released because of her truculent tendencies. This dismissal marked an early indication of the turbulent path her Hollywood career would take.

A Hit in Hollywood

Despite her belligerent reputation, or perhaps partly because of it, the burgeoning movie industry took note of the young and talented Brooks, casting her in a slew of silent films. She was popular with the public and quickly became the premiere on-screen darling of Paramount. Offscreen, though, her bosses might have described her as anything but darling.

Only a couple of years into her film career, Louise was already at the top of the silent film heap; but, like F. Scott Fitzgerald's description of his wife, Zelda, Louise, too, thought of herself as a failed social creature. This weakness was her undoing. Unable—or at least unwilling—to comply with societal niceties, Louise was revered for her beauty, as well as for a charisma that translated effervescently to the big screen, but her talents were overshadowed by her unabashed honesty. Although Paramount capitalized on her acting aptitude (and Hollywood men on her sexual liberty), Louise irrecoverably cut ties with Paramount when executives reneged on one too many offers.

Europe Beckons

While on leave from Hollywood, Louise was quickly snatched up by German Expressionist filmmaker G. W. Pabst and was cast as Lulu in *Pandora's Box*. This now-infamous character was a sexually charged vaudevillian whose amorous exploits included the first-ever on-screen depiction of lesbian relations. Finally, a genre that seemed fitting to Louise's sexually charged and controversial nature! On the heels of *Pandora's Box* came *Diary of a Lost Girl*, another Pabst-directed exposé on society's sexual underbelly and social conduct. In this film, Louise's character, Thymiane, suffers through rape, the birth of her illegitimate child, the death of said child, and eventually, prostitution.

No Chance for a Comeback

Upon her return to Hollywood in 1930, Louise fell victim to the cold-hearted punishment of blacklisting. Unhappy with Paramount, she had refused to participate in sound editing for *The Canary Murder Case*. Thus, the film showcased Louise's role as "The Canary" with another songbird, Margaret Livingston, on vocals. Because her prior films had all been silent, Paramount was able to claim that Louise's melodious voice was inept for singing in talkies, a low blow that solidified her downhill reputation. From then on, Louise moved from roles in which she was critically ignored to low-budget films to an embarrassing final role in a John Wayne Western.

Louise never made it back to her short-lived Hollywood high point. She did, however, make a blaring reappearance late in life through a writing career in which she brutally dissected the world of cinema. In Louise's own words, the character of Lulu "dropped dead in an acute attack of indigestion" after devouring her "sex victims." It appears that fickle Hollywood had chewed up and spit out young Louise.

The Mondo Movie Craze

This movie trend in the 1960s was all about celebrating excess for its own sake. How far would they go?

With a warning that reads in part: "The duty of the chronicler is not to sweeten the truth but to report it objectively," *Mondo Cane (A Dog's World)*, the first mainstream "shockumentary," unspooled before thrill-seeking American audiences in 1963. The film promised to "enter a hundred incredible worlds where the camera has never gone before" and presented a loosely strung travelogue of outrageous rites, repulsive rituals, and downright bizarre behavior.

More, More, More

A combination of archival footage and staged sequences shot in grainy documentary style to appear authentic, *Mondo Cane* featured a wide range of shocking footage—including mass animal slaughter, a group of religious women tongue-bathing parish steps, a visit to a pet cemetery, diners feasting on cooked insects, a match in which two natives conk each other on the head with logs, and plenty of topless African women.

The 108-minute film, underscored by sardonic narration, weaves back and forth between the "primitive" and the "civilized" world. The outrageousness is further accentuated by the inclusion of a perky song titled "More" that is endlessly repeated (the song garnered an Oscar nomination for Best Song in 1963).

Mondo Cane was cowritten and directed by Paolo Cavara and Gualtiero Jacopeti. The word mondo quickly became a euphemism for "extreme." The movie was so popular around the world that it started a trend of similar films throughout the 1960s. Each subsequent mondo release attempted to top the previous one. Titles included *Mondo Macabro, Mondo Mod, Mondo Exotica,* and *Mondo Topless.* By the time *Mondo Bizarro* was released in 1966, the footage in the films was almost all staged. The genre petered out by the end of the decade, but a resurgence of the genre occurred in the late 1970s with the *Faces of Death* series, whose hallmark was explicit and gory sequences. The influence of the mondo films can be seen in recent reality TV shows such as *Fear Factor and Survivor.*

Welcome to the Kalakuta Republic

Fela Kuti is an internationally known Afropop musician. He was also the leader of a Nigerian anti-government movement. Read on to discover one man's trials and tribulations, as seen from his compound, otherwise known as the "Kalakuta Republic."

A Corrupt Country

In the mid-1970s, Nigeria, with more than 100 million citizens, was the most populous country in Africa; as a new member of the Organization of Petroleum Exporting Countries (OPEC), it was also one of the world's leading oil exporters. Although enormous amounts of foreign money flowed into the government's coffers, it quickly flowed into the hands of the ruling class. But life for the vast majority of Nigerians was as hard as ever. Crime was rampant, poverty widespread, and all forms of dissention were violently suppressed. Yet one man in Lagos continued to publicly criticize the rulers and their backers: the musician, Fela Kuti.

From Lagos to London to Los Angeles

Afropop is a hypnotic, compelling blend of American funk arrangements, European classical compositional technique, jazz, and African tribal rhythms. Kuti, the genre's originator, was the son of a decidedly left-wing, middle-class family from Lagos. His jazz highlife band, Koola Lobitos, attracted sufficient attention to justify a 1969 tour of the United States. Kuti would later say that the 10 months he spent in America galvanized all of his later political thinking.

In the United States, Kuti was exposed to the black power movement. A Black Panther friend gave him a copy of Malcolm X's autobiography; Kuti was an instant convert. He returned to Nigeria and formed Africa 70, a huge ensemble complete with horns, saxophonists, guitarists, dancers, and singers. Kuti and his band established themselves at the local club Shrine.

Although his songs frequently attacked the status quo, the country's leaders were generally content to leave the "crazy" musician alone.

An Independent Republic

Kuti established the area around his Lagos home as a commune, complete with medical facilities, farm animals, and a recording studio. Kuti went a step further and had the compound fenced with electrified barbed wire and proclaimed it a sovereign nation—the Kalakuta Republic. With Africa 70 selling millions of albums, Kuti's political and social diatribes became increasingly brazen. Military leaders hated him for his flagrant criticism, and the bourgeoisie hated him because of his free lifestyle.

Tensions between Kuti and the military reached a breaking point in 1976, following the second World Black and African Festival of the Arts and Culture. Kuti withdrew Africa 70 from the official lineup and staged a counter festival at Shrine, where he debuted "Zombie," a song mocking government soldiers. "Zombie" became an overnight sensation, but for the military, it was the last straw.

The Raid and Its Aftermath

On February 18, 1977, more than 1,000 soldiers amassed outside the Kalakuta Republic compound. They barricaded the building, set fire to the generator, and attacked the building with savage ferocity. Soldiers beat people, raped women, and smashed equipment. Kuti's aging mother, herself a renowned political activist, was thrown through a window and later died from her injuries. Kuti was dragged from the building and severely beaten, suffering a fractured skull and broken bones. The rest of the residents were carted to jail or the hospital. A government-sponsored committee later found no wrongdoing on the part of the soldiers.

Kuti and his music survived the raid. He established a new Kalakuta Republic and one year later married 27 women in a Yoruba ceremony. Though he was often jailed and beaten, he continued to reside in Lagos where he performed and preached against the ruling power. Kuti's persistence earned him the name *Abami Eda*, or "Chief Priest," among his fans. He remained a vibrant force in Nigeria and in world music until his death from AIDS in 1997. More than one million people attended his funeral procession.

The Strange World of Joe Meek

The music business has produced more than its fair share of eccentric characters, but very few have been stranger than Joe Meek, the independent British pop record producer of the 1950s and '60s.

A Complicated Man

Perhaps you have never heard of Joe Meek, but you've probably heard the musical recordings he produced and inspired: Meek was the first to develop the electronic effect reverb and sampling techniques. In 1962, he produced "Telstar," an electronics-soaked single by the Tornados that became a worldwide sensation. It was the first record by a British group to top the charts in the United States, more than a year before the arrival of The Beatles. "Telstar" sold more than five million copies around the world, making it one of the most successful instrumental records ever. Meek also recorded with then-unknown singer Tom Jones, although it was another two years (and with a different producer) before the singer made it big.

But that wasn't all there was to Meek. As well as being an innovative music producer, he was also an occultist, a paranoid eccentric, and, illegally at the time, a homosexual. Despite Meek's successes and influence on modern music, it's his strangeness that people remember.

304 Holloway Road

Joe Meek created his legendary home studio in a three-story apartment above a leather goods store in London. Meek's landlady operated the store below and frequently complained about the noise her tenant made. In the days before eight-track recording, records were recorded live with dubs added later. To obtain just the sound he was looking for, Meek would often have performers playing on the stairs or in the bathroom. When his long-suffering landlady complained by banging on the ceiling, Meek would simply turn up the volume to drown her out.

Meek's Downward Spiral

An obsessive Buddy Holly fan, Meek believed that the late rocker communicated with him from beyond the grave. He developed a similar obsession with the still-living bass player from the Tornados, Heinz Burt. When Meek confiscated a shotgun that Burt had brought to the studio one day, the components for violence were all in place.

In 1963, Meek was arrested in a men's public restroom after allegedly "smiling at an old man," which made the national press. Meek spiraled into depression, and his paranoia and outbursts worsened. On February 3, 1967, after yet another quarrel with his landlady, Meek blasted her with the shotgun that he had confiscated from Burt. He then turned the gun on himself. Coincidentally, it was also the eighth anniversary of Buddy Holly's death.

• Like many musicians at the time, Meek was known to pop plenty of pills and experiment with LSD. The drugs may have accounted for much of his strange behavior.

• Meek was once seen running alone late at night along London's Holloway Road, dressed in pajamas and screaming that he was being chased by someone with a knife.

• When Phil Spector, a legendary (and legendarily whacked-out) producer in his own right called Meek to tell him how much he loved his music, Meek accused Spector of stealing his ideas and hung up on him.

• Buddy Holly wasn't the only dead guy talking to Meek—he also believed he was in spiritual contact with the ancient Egyptian Emperor Rameses the Great.

The Weird, Wacky, and Wonderful *Wizard of Oz*

The stories behind the 1939 classic are almost as entertaining as the movie itself.

The Story

When L. Frank Baum wrote the book *The Wonderful Wizard of Oz* back in 1900, he had no idea what he was starting. The story follows the adventures of Dorothy Gale, a young girl from Kansas, who, along with her dog, Toto, gets caught up in a tornado and is magically transported to the land of Oz. The book sold tens of thousands of copies to a public who ate up the strange and wonderful tale.

In 1938, MGM bought the rights to the novel and adapted it for film. Many details were changed, including the famous shoes Dorothy wears: in the book, they were silver, not red. But the major difference between the movie and the book is that Baum's Dorothy really does go to a place named Oz; MGM felt that making Oz a place Dorothy visits in a dream would better explain the psychological motivations of her desire for love and acceptance. So, after countless revisions, the script was finalized and production began.

The Cast

The cast you know and love from the movie wasn't the cast that producers started with. Many, many recasts were made from preproduction until after filming began. Shirley Temple was rumored to be up for the part of Dorothy, and Buddy Ebsen was the original Tin Man. But after suffering an allergic reaction to the silver paint used in the makeup, he was admitted to the hospital, and the role went to Jack Haley instead. Haley was unaware of what caused Ebsen's illness, and the makeup formula was changed to avoid a second disaster. The role of the Wicked Witch originally belonged to Gale Sondergaard, but she quit after execs changed the character from a haughty, glamorous witch to a green-faced old hag. Margaret Hamilton replaced her. And W. C. Fields was asked to play the Wizard, but he reportedly had a scheduling conflict with *You Can't Cheat an Honest Man*, a vehicle designed especially for his talents and thus a better opportunity for him. He was replaced by Frank Morgan.

Once the casting had settled down, the studio likely thought the snags in production were over. Not quite.

The Trouble

Only a few weeks into filming, director Richard Thorpe was fired by MGM. He was replaced by George Cukor, who would soon take on directing *Gone with the Wind* (1939). Cukor never actually filmed any scenes for *Oz* however; he was replaced by Victor Fleming, who was later joined by director King Vidor, who shot the black-and-white scenes. (And you thought the casting was complicated.)

In addition to personnel issues, physical injuries also ran rampant on the set. There was the incident with Ebsen and the makeup, and his replacement, Jack Haley, also endured a serious eye infection because of the makeup. Then, one day while filming, Margaret Hamilton was severely burned during a pyrotechnics accident. When she wasn't recuperating from burns, she was trying not to ingest any of the toxic makeup used to create her witch persona.

And then there were accounts from Judy Garland later in life about the pills administered to her by the studio to keep her weight down. Apparently, the studio plied her with amphetamines to control her weight and increase her productivity. As a result, she battled an addiction to pills the rest of her life, only to have MGM drop her contract during the 1950s because the addiction made her unreliable. In 1969, Garland died of an accidental overdose.

Moreover, the directors, producers, studio heads, songwriters, and scriptwriters all had opinions about the direction of the story, which slowed production and fired up more than a few tempers. Originally, there was a singing contest number in Oz in which Dorothy's "hip" vocal style won the hearts of the Munchkins. A group number known as "The Jitterbug" was taken out of the movie after five weeks and thousands of dollars were spent filming it, and, astonishingly, "Over the Rainbow," Dorothy's sweet, sad song (ranked No. 1 on the American Film Institute's list of 100 Greatest Songs in American Films), was also cut from the picture because the studio thought the film was running too long. Producer Arthur Freed and lyricist Harold Arlen, convinced of the song's power, lobbied hard for MGM founder Louis B. Mayer to put it back in. He begrudgingly agreed, which was a smart move because it won the Academy Award for Best Song. The movie itself won an Oscar for Best Original Score and was nominated for Best Picture as well as three technical awards.

The Reception

The film that cost MGM $2.8 million to produce was released in 1939 and made $3 million in its first theatrical run. Because of advertising and distribution costs, that wasn't enough of a profit for the studio to consider it a success, so ten years later, the film was rereleased and earned an additional $1.5 million, which made the producers happy.

But *The Wizard of Oz* is often considered "the movie that television made." In 1956, CBS broadcast the movie to an audience of 45 million viewers—even though most people at the time didn't have color TV sets and therefore never saw the movie's sequences in Technicolor. Three years later, the movie was broadcast again as a two-hour Christmas special and the response was overwhelming. People loved *The Wizard of Oz* so much that the movie was broadcast every year for decades, traditionally around special holidays, first by CBS, then by NBC. You can still catch it on cable channels or just reach into your video or DVD library. *The Wizard of Oz* has enjoyed some of the highest home video/DVD sales in history—millions of people own a copy of one of the many editions available for home viewing.

Fast Facts: *Star Wars*

• Released in 1977, the first *Star Wars* movie cost just over $11 million to make, compared to $113 million for *Star Wars: Episode III—Revenge of the Sith* (2005), which was produced nearly 30 years later.

• In the late 1990s, remastering and reediting *Star Wars* for its 20th anniversary edition cost nearly as much as it did to make the original movie.

- The first U.S. theater run for the original *Star Wars* pulled in $215 million.

- The first-ever *Star Wars* trailer began showing a full six months before the movie came out. Vague taglines such as "the story of a boy, a girl, and a universe" and "a billion years in the making" were meant to build up buzz before the film's debut.

- The first *Star Wars* film was originally going to be titled The Star Wars, but George Lucas decided to drop the introductory article. The movie's full title, *Star Wars: Episode IV: A New Hope*, was not used on posters, promotions, or publicity until the film was rereleased in 1981.

- George Lucas was initially set to receive about $165,000 for the making of *Star Wars*. But when production costs rose, he waived his fee in exchange for 40 percent of the box-office returns. That, combined with a lucrative merchandising deal and his foresight into snagging creative control and full merchandising rights for the sequels, made him millions.

- The filmmaker who directed *Scarface* (1983) cowrote the opening crawl text at the beginning of Star Wars. Brian De Palma helped pen the words: "It is a period of civil war. Rebel spaceships, striking from a hidden base, have won their first victory against the evil Galactic Empire . . ."

- The actors who played C-3PO and R2-D2—Anthony Daniels and Kenny Baker, respectively—are the only two people credited with appearing in all of the first six *Star Wars* movies.

- Daniels' C-3PO costume was precision engineered and fit very tightly. If he moved too much, the pieces of the suit cut into him. Walking in the costume resulted in a great many scrapes, cuts, and abrasions—a problem throughout the shoot.

CULTURAL CURIOSITIES

- After Harrison Ford tested for the part of Han Solo, he had the edge for the role, though early in the project, Lucas had decided he didn't want to use anyone he had directed in the past. The potential candidate list included Kurt Russell and Christopher Walken, among others.

- Jodie Foster and Cindy Williams auditioned for the role of Princess Leia, along with Linda Purl, Terry Nunn, and many others. Foster and Nunn were rejected because they were under 18 at the time.

- The character of Luke Skywalker went through many incarnations during script development. In the original treatment for the *Star Wars* script, which was written in 1973, Luke Skywalker was a general assigned to protect a rebel princess. In a later version, his name was Kane Starkiller, and he was a half-man, half-machine character who is friends with Han Solo. Later, he evolved into the young man closer to the character we are familiar with, but his name was Luke Starkiller. Just before shooting began, Lucas decided that the name might remind people of murderers like Charles Manson, so the name was changed to Luke Skywalker.

- Several desert locales on Luke Skywalker's home planet of Tatooine are in Death Valley National Park, California. Several other Tatooine locations—including Luke's home—are in Tunisia.

- The stormtroopers' weapons in *Star Wars* were retooled military weapons from the 1940s.

- In the final Luke-versus-Vader battle sequence of *The Empire Strikes Back*, Luke kicks the Sith lord off a ledge in a carbonite chamber, then jumps down after him. Just before the scene cuts, Luke's head reappears as the actor bounces back up from the trampoline he landed on.

- The first kiss between Princess Leia and Han Solo was a long time coming, and that made it one of Hollywood's most memorable. Fisher's character says at one point that she'd "rather kiss a Wookie" than the handsome pilot played by Ford, but that only added to the audience's anticipation. When the two finally lip-locked, it was well worth the *Star Wars*-size wait.

Krypton, U.S.A.!

According to the story, Superman came to Earth from the planet Krypton. But in reality, the Man of Steel is a native of Cleveland.

Jerry Siegel and Joe Shuster were pals at Glenville High School who shared a fondness for comics, adventure movies, and pulp magazines. Their dream was to make it big in the burgeoning field of comics, and in their early 20s, they managed to sell a few detective and adventure stories to Malcolm Wheeler-Nicholson, one of the first comic book publishers to use original material. Their greatest success, however, would come years later in the form of a man with a big red S on his chest.

A Superhero Is Born

Superman was an amalgam of concepts and images that had burned themselves into Siegel's young brain, including body-building ads in the backs of magazines, the popular pulp hero Doc Savage, and Philip Wylie's influential science fiction novel, *Gladiator*. Siegel distilled it all into a costumed superhero like no other, an almost indestructible figure who used his remarkable strength to fight for truth and justice.

Siegel and Shuster initially tried to sell Superman as a newspaper strip but were rejected by every syndicate they approached. They fared no better with various comic book publishers until Superman landed in the hands of Sheldon Mayer, an editor at what later became DC Comics, who sensed a potential appeal to teenage boys in the primitive strip. Superman was picked for the lead feature for the company's newest title, *Action Comics*, and the two were paid $130 for all rights to the character.

Supermania Sets In

Action Comics #1 hit the stands in the summer of 1938 and was an immediate success. Kids couldn't get enough of the Man of Steel, and he was quickly given his own title. The first issue sold 900,000 copies. Supermania swept the nation as the character appeared on everything from lunch boxes to decoder rings

Siegel and Shuster were making pretty good money writing and illustrating Superman's adventures, but because they had sold all rights to their creation, they shared none of the merchandising profits. They took DC Comics to court in 1947 in an attempt to get a percentage of the revenues being generated by their character. The courts ruled in the publisher's favor, and the two soon found themselves out of a job. Over the years, they found piecemeal work for DC and other publishers, but Shuster eventually developed vision problems that prevented him from drawing, and Siegel was forced to take a $7,000-a-year civil service job for the health benefits. By the 1970s, both men were practically destitute.

Righting Wrongs

In 1975, the press became aware of how shabbily the creators of Superman had been treated. The National Cartoonists Society took up their cause, enlisting the support of influential writers to force DC Comics to reward the men who had made the company so much money. After months of hard-fought negotiations, DC Comics agreed to a settlement that included $20,000 a year for each creator with built-in cost-of-living increases, provisions for their heirs, and, most importantly, creator credit on almost everything on which Superman appears.

The Man of Steel Today

Since his debut in 1938, Superman has become one of the most recognizable fictional characters in the world. He's saved Earth countless times, fought outrageous villains and monsters, and even come back from the dead. Outside of comic books, he has been featured in serials, television series, motion pictures, animated cartoon series, and video games. Per their deal with DC Comics, all of the above are noted as "created by Jerry Siegel and Joe Shuster." Not bad for two Cleveland kids with a supersize dream.

The Great Ghoulardi

Long before Howard Stern, a different type of shock jock stirred up America's airwaves. Ghoulardi, a character created by actor and announcer Ernie Anderson, ruffled plenty of feathers during his days on Cleveland's WJW-TV. His unique antics still live on in memories and pop culture tributes today.

To many, it may sound like a kind of gourmet chocolate, but for families in Cleveland, the name Ghoulardi is anything but sweet. Ghoulardi was the brainchild of WJW-TV announcer Ernie Anderson, who created the character in 1963 as part of his hosting role on WJW's *Shock Theater* program—and shock is definitely what he did.

Meet Ghoulardi

Ghoulardi likely had a tough time blending into a crowd. The character was known for his outlandish outfits, which featured zany wigs, a long lab coat covered in colorful buttons, and his trademark sunglasses with only one lens intact. Ghoulardi also had a stylish fake goatee. The name was said to be inspired by the word "ghoul" and the last name of a local make-up artist, Ralph Gulko.

The greatest memories of Ghoulardi, however, may be the things he said: The unusual fellow flung catchphrase after catchphrase into his chatter. One minute, he might be saying "turn blue," while the next, he'd be talking about "knifs" (a "knif " being "fink" spelled backwards).

Ghoulardi would appear during breaks in various horror films shown during the *Shock Theater* show. He was known for telling the audience how terrible a particular flick was, or suggesting they might be better off going to bed rather than staying up for the show. (The program aired late on Friday nights.) He even went so far as to pop random, non sequitur clips in the middle of the movies, often featuring either himself or just irrelevant stock footage. The bits regularly featured memorable music, too: Anderson often worked little-known "hip" tunes into the background during his various segments.

Successes and Stresses

Ghoulardi was without question a hit. The TV station sold merchandise based on the character, and a charity sports team was even founded under the name. Still, tensions exist-

ed between Anderson and the station. Then-extreme tactics such as setting off firecrackers in the studio were said to have made executives nervous. It didn't help that the character constantly ridiculed Cleveland's suburbs, most notably the area of Parma. Skits branded as "Parma Place" mocked the working class neighborhood to no end.

Ghoulardi stayed on the air until the end of 1966, when Anderson decided to retire the character to pursue an acting career in Los Angeles.

After Ghoulardi

Anderson's acting career never materialized, but that doesn't mean he didn't succeed in Hollywood: After leaving Ghoulardi behind, Anderson landed a gig as the main voiceover talent for ABC. He held the role for about two decades.

In 1997, Anderson died following a battle with cancer. His legacy went on, though: One of his former cohorts, an intern named Ron Sweed, started a program called *The Ghoulardi Show* that was based on the Ghoulardi character. It went on to receive nationwide syndication. Comedian and Cleveland native Drew Carey honored the Ghoulardi reputation by wearing a Ghoulardi shirt in episodes of *The Drew Carey Show* and offering a quote in a book about the character. What might be the most touching tribute, though, comes from right within

Anderson's own family: His son, film director Paul Thomas Anderson (*Boogie Nights, There Will Be Blood*) named his production unit "The Ghoulardi Film Company."

Stars Behind Bars!

Most actors have done their time working their way up the ladder through roles in B-movies, television, or theater. But a surprising number of actors have literally done time—as in prison time. Here's a sample:

Lillo Brancato

Brancato played Robert De Niro's son in *A Bronx Tale* (1993) and a bumbling mobster on *The Sopranos*. But drug addiction took its toll on his career. In December 2005, Brancato and a friend broke into an apartment looking for drugs. In the process, an off-duty policeman was shot and killed. Brancato was charged with second-degree murder and attempted burglary. He was acquitted on the murder charges in 2008 but served time for attempted burglary.

Errol Flynn

One of the most popular leading men in Hollywood history, Errol Flynn frequently found himself in trouble with the law. His various stints behind bars included two weeks in a New Guinea jail for hitting an Asian man who addressed him without the prefix "Mr." and several days in lock-up for striking a customs officer in the tiny African country of Djibouti.

Stacy Keach

In the mid-1980s, the star of the acclaimed Western *The Long Riders* (1980) served six months in prison for smuggling cocaine into England.

Paul Kelly

Paul Kelly played lead roles in many B-films, mostly crime melodramas. In the late 1920s, he killed his best friend, actor Ray Raymond, in a fistfight over Raymond's wife, actress Dorothy MacKaye. He served two years for manslaughter, then went on to a successful film and stage career.

Robert Mitchum

As a teenager in the 1930s, Mitchum was arrested for vagrancy in Georgia and was sentenced to a week's work on a chain gang, but he escaped the first chance he got. In California, in 1948, the movie tough guy served 50 days in jail for marijuana possession.

Tommy Rettig

As a child actor, Rettig gained lasting fame as Lassie's master in the popular 1950s TV series. But in 1972, he was arrested for growing marijuana, and in the mid-1970s, he was sentenced to five-and-a-half-years in prison for smuggling cocaine into the U.S. The charges were dropped after an appeal, as was another drug charge five years later.

O. J. Simpson

A football Hall of Famer whose movies include *Capricorn One* (1978) and *The Naked Gun* (1988), Simpson was acquitted of the murder of his ex-wife, Nicole Brown, in 1995. In December

2008, the football superstar was handed a sentence of 9 to 33 years in prison for armed robbery and kidnapping as a result of a botched attempt to get back items Simpson claimed a sports memorabilia dealer had stolen from him.

Christian Slater
In 1989, Slater was involved in a drunken car chase that ended when he crashed into a telephone pole and kicked a policeman while trying to escape. He was charged with evading police, driving under the influence, assault with a deadly weapon (his boots), and driving with a suspended license. In 1994, Slater was arrested for trying to bring a gun onto a plane. In 1997, he was sentenced to 90 days in jail for cocaine abuse, battery, and assault with a deadly weapon.

Mae West
In 1926, Mae West, one of Hollywood's most iconic sex symbols, was sentenced to ten days in jail when her Broadway show, *Sex*, was declared obscene.

6 Stars Who Turned Down Memorable Roles

It seems that the actors cast in our favorite movies are perfect for the part, but they're often not the director's first choice. Most celebs have turned down more roles than they've taken, some with regrets but many with thanks to their lucky stars.

1. Brad Pitt has had hunky roles in many films including *Troy* (2004), *Legends of the Fall* (1994), and *Thelma & Louise* (1991). But he turned down a role in *Apollo 13* (1995) to make the movie *Se7en* (1995), which won an MTV Movie Award for Best Movie, beating out *Apollo 13*. But *Apollo 13* received nine Academy Award nominations and won two Oscars, leaving *Se7en* in the dust.

2. **Mel Gibson** has had a blockbuster career as an actor, starring in both the *Mad Max* and *Lethal Weapon* series, and as a director, winning an Academy Award for *Braveheart* (1995), in which he also starred. Gibson turned down the lead role in *The Terminator* (1984), which went to Arnold Schwarzenegger instead. Gibson was also offered the lead in the first *Batman* movie (1989) (which went to Michael Keaton), but he was already committed to *Lethal Weapon 2* (1989). Later, he turned down the part of villain Two-Face in *Batman Forever* (1995), which went to Tommy Lee Jones.

3. A star with a career as long as **Sean Connery's** is bound to make both good and bad decisions. The good ones include roles in *Indiana Jones and the Last Crusade* (1989), *The Hunt for Red October* (1990), seven James Bond movies, and his Academy Award-winning role in *The Untouchables* (1987). But one questionable decision was turning down the 007 role in *Live and Let Die* (1973), which became a great career move for his replacement, Roger Moore. Connery later turned down the role of Gandalf in the *Lord of the Rings* trilogy, which went to Ian McKellen, and the role of Morpheus in *The Matrix* films, which went to Laurence Fishburne—two decisions that he admitted regretting.

4. **Al Pacino** rose to fame playing Michael Corleone in *The Godfather* movies and has since starred in many great movies, including *Scarface* (1983), *Donnie Brasco* (1997), and *Scent of a Woman* (1992), for which he won an Oscar. But can you imagine Big Al as Han Solo in *Star Wars* (1977), the role that started the career of Harrison Ford? Pacino also turned down the lead role in *Close Encounters of the Third Kind* (1977), which instead went to Richard Dreyfuss. But the actor Pacino has had the closest encounters with is Dustin Hoffman. Pacino turned down starring roles in *Midnight Cowboy* (1969), *Marathon Man* (1976), and *Kramer vs. Kramer* (1979), all of which went to Hoffman.

5. Rock Hudson, a favorite leading man of the 1950s and 1960s, starred in films such as *Come September* (1961), *Oz* (1964), and *Pillow Talk* (1959), the first of several films that costarred Doris Day. He signed on to play the lead in *Ben Hur* (1959), but when contract negotiations broke down, the part went to Charlton Heston instead, an outcome that would ultimately be Hudson's only career regret.

6. Will Smith began his showbiz career as half of the hip-hop duo DJ Jazzy Jeff & the Fresh Prince, who won the first ever Grammy in the Rap category in 1988. Smith was nearly bankrupt by 1990, when he was hired to star in the sitcom *The Fresh Prince of Bel-Air*, which became a huge success. His movies include *Independence Day* (1996), *Men in Black* (1997), and *Ali* (2001). He was also offered the lead role in *The Matrix* (1999). Despite the film's phenomenal success, Smith later said that he didn't regret turning down the role because Keanu Reeves "was brilliant as Neo." Smith also originally passed on *Men in Black*, but his wife convinced him to reconsider.

CHAPTER 5

WAR AND PEACE

23 Military Awards Won by Audie Murphy

Audie Murphy was the most decorated U.S. soldier during World War II. He received 23 awards, including 5 awards from France and Belgium, and every decoration of valor that the United States offered. After the war, he pursued a successful movie career, starring in films such as *The Red Badge of Courage, The Unforgiven, and To Hell and Back*, based on his autobiography. Murphy won each of the following 23 awards—some of them more than once.

1. Medal of Honor
2. Distinguished Service Cross
3. Silver Star with First Oak Leaf Cluster
4. Legion of Merit
5. Bronze Star with V Device and First Oak Leaf Cluster
6. Purple Heart with Second Oak Leaf Cluster
7. U.S. Army Outstanding Civilian Service Medal
8. Good Conduct Medal
9. Distinguished Unit Emblem with First Oak Leaf Cluster
10. American Campaign Medal
11. European-African-Middle Eastern Campaign Medal
12. World War II Victory Medal
13. Army of Occupation with Germany Clasp
14. Armed Forces Reserve Medal
15. Combat Infantry Badge
16. Marksman Badge with Rifle Bar
17. Expert Badge with Bayonet Bar
18. French Fourragere

19. French Legion of Honor, Grade of Chevalier
20. French Croix de Guerre with Silver Star
21. French Croix de Guerre with Palm
22. Medal of Liberated France
23. Belgian Croix de Guerre 1940 Palm

Armor: Outstanding and Interesting Tanks

The twentieth century saw the armored fighting vehicle (AFV) supplant heavy cavalry as a swift, hard-hitting arm of war.

Some tanks have been a cut above their contemporaries, while others were funny looking, or failures, or both. Let's start with the great ones.

The Great

FT-17 (France, 1917; 37mm or machine gun): Unlike most elephantine World War I armor, the small FT-17 had a 360-degree rotating gun turret, and it crossed trenches with the help of extended rear rockers rather than oversized, diamond-shaped treads. It was the first World War I tank that looked like World War II tanks, and it was the only one to see significant World War II service.

Matilda II (U.K., 1937; 40mm): If you want to stop the enemy in its tracks, bring a tank that can barely be harmed. Mattie's crews could laugh at German tank cannons until 1942, when most Soviet Matildas were phased out in favor of faster, more powerful tanks. Although the relatively small Matilda waltzed slowly, her bulldog presence commanded respect.

KV-1 (U.S.S.R., 1939; 76mm): This hulking brute owes its fame partly to timing. When Germany invaded the Soviet Union, the KV was a monster only the Stuka dive-bomber or 88mm flak gun could slay. Outside Leningrad, one KV-1 withstood 135 German cannon hits.

T-34 (U.S.S.R., 1941; 76mm, then 85mm): Looking for credibility around World War II zealots? Call this the best tank of the war. Its amazing speed, sloped armor, and strong gunnery enabled aggressive tank tactics, perfectly suited to Eastern Front warfare and deadly to Germany. The 1944 T-34 (85) model was still serving in some armies in 2000.

Panzerkampfwagen V Panther (Germany, 1943; 75mm): Remember the guys who sneered when you heaped praise on the T-34? They have nothing on Panther advocates. Speedy and well armored up front, the Panther's hard-hitting long gun could engage at ranges that enabled few enemies to harm it.

M-60 Patton (U.S., 1960; 105mm): For two decades, this reliable tank was NATO's mainstay—and perhaps the finest tank of the 1960s. Its variants formed the backbone of Israeli armor during the Yom Kippur War, and the U.S. Marines drove some into battle during the 1991 Gulf War. The trustworthy M-60s were routinely updated until the end of the Cold War.
T-72 (U.S.S.R., 1971; 125mm): Any tank this prolific rates mention. When introduced, the T-72's 125mm cannon raised the gunnery bar above the common 105mm. It proved fast and reliable, while delivering a lot of bang for the ruble, and in 2002 it was still the world's most widely deployed tank.

The Odd

A7V (Germany, 1917; 57mm): This clunker looked like a railroad caboose covered with a big steel drop cloth, and it fought

War and Peace

about as well. It couldn't cross trenches, which was disappointing, because the goal of tanks was to break up trench warfare stalemates. If you romanticize German tanks, try not to look too hard at pictures of this armored banana slug.

Fiat 2000 (Italy, 1917; 65mm): While heavily armored and bristling with seven machine guns, the slow F-2000 could barely outrun a briskly marching infantryman. Fortunately for its crews, it never had to try; it only served in peacetime.

M13/40 (Italy, 1940; 47mm): Lousy armor, weak gun, underpowered and unreliable engine, prone to stalling or catching fire when hit—what a combination. This bowser was the mainstay of Italian armor in North Africa. With equipment like this, who can blame Italian crews for bailing out of their tanks and surrendering?

Elefant tank destroyer (Germany, 1943; 88mm): A perennial candidate for Dumbest World War II Armor Design, the glacially slow Elefant mounted the famous 88mm cannon and looked like today's self-propelled howitzers. It had no machine guns of its own, so enemy infantry was welcome to spray tags on it or use its massive front housing as a latrine.

Churchill Crocodile (U.K., 1944; 75mm, flamethrower): Most World War II powers developed flamethrowing tanks; this was one of the best designs. The Crocodile pulled a trailer of modified gasoline "ammo" yet still mounted a standard tank cannon. Hosing flaming gas the length of a football field, the Croc was hell for dug-in defenders.

Sherman DD (U.S./U.K., 1944; 75mm): Tanks don't swim well, which is an issue when you're planning a D-Day. Allied engineers invented a watertight skirting for the Sherman so that it could float, and then added a little propeller for instant buoyancy—at least enough for the tank to reach the beach.

Ten Sherman DD battalions, the equivalent of a full division, waded ashore this way on June 6, 1944. A number swamped and sank, but on that day any Allied tank on shore was very welcome.

The Death of Davy Crockett

You can start fights in Texas suggesting that the Alamo fighters—Davy Crockett was prominent among them—didn't die fighting to the last man. But do we really know that they did?

First, the orthodox version: Davy Crockett grew up rough and ready in Tennessee, wrestling bears and otherwise demonstrating his machismo. He went into politics, lost an election, and moved to Texas. His homespun, informal braggadocio went over just fine in what would soon become the self-proclaimed Republic of Texas. Mexico, of course, didn't grant that Texas had the right to secede. General Antonio López de Santa Anna invaded the Republic and cornered one group of its defenders in Alamo Mission, San Antonio de Béxar (now just San Antonio). Shortly before independence, Crockett had signed on to fight for Texas, and he was one of the men who died fighting in its defense.

Why question that? Most of the questions stem from the memoir of a Mexican officer who fought there, Jóse Enrique de la Peña. De la Peña says that seven captives, including Crockett, were brought before Santa Anna and murdered in cold blood after the battle. He also says that Crockett took refuge in the Alamo as a neutral foreigner rather than as a volunteer militia-man. That poses authenticity problems, because there was no logical reason for Crockett to be at the Alamo unless he planned to fight in its defense. It's worth noting that de la Peña says he found the execution appalling. However, the Mexican officer also tells an implausible version of the death of Colonel William Travis, commander in charge of the siege of the Alamo: He claims to have seen it occur, but there's little chance de la Peña could have positively identified Travis at a distance. This was, after all, a battle with thick black-powder smoke, hand-to-hand combat, and concealment.

What evidence supports the orthodox version? Travis's slave, Joe, survived the battle and says Travis died defending the north wall (not where de la Peña has him). Joe also says he saw Crockett's body surrounded by dead Mexican soldiers, and an officer's wife who survived also testifies that Crockett died in battle. Santa Anna himself didn't say anything about executing Crockett in his after-action report; he did say that Crockett's body was found along with those of other leaders.

Why is it contentious? The memory of the Alamo is a Texan cultural rallying point. Opposing this view is a revisionist stance that seems so ready to dismiss tales of military valor that it dumps the orthodox account as too simple and perfect to be true. Either side has generally drawn a conclusion and seeks evidence to support it.

Do we know? So much happened at Alamo Mission between February 23 and March 6, 1836, that we will never know. It is plausible that some wounded survivors, possibly dying, were executed after the battle; that doesn't negate anyone's heroism. What's lacking is compelling, credible evidence to contradict the eyewitnesses who report no such thing. Absent that evidence, and with de la Peña's writing a questionable account well after the fact, the weight of documentation suggests that Davy Crockett went down fighting.

Pappy Boyington and His Black Sheep

The men of VMF-214 had a reputation as the black sheep of the U.S. Air Corps. However, the tactics and skill of the pilots and their commanding officer, "Pappy" Boyington, made them heroes.

It was August 4, 1941, and Marine Corps aviator Gregory Boyington had reached rock bottom. Stationed in Pensacola, Florida, he was broke, his wife and children had left him, and his reputation for brawling and drinking had eliminated his chances of promotion despite his talent as a pilot.

That night in Florida, he knew that a representative of the Central Air Manufacturing Company was in a nearby hotel recruiting pilots for a volunteer mission in China. Tired of the Marines and lured by the promise of money, Boyington stopped at the bar for a few drinks and signed up.

He trained with the American Volunteer Group (AVG) in China, also known as the Flying Tigers. Led by Claire L. Chennault, the notorious group learned tactics to combat Japan's best pilots. Boyington remained a drinker and brawler, but with six kills, he earned another sort of reputation as a formidable combat pilot. However, Chennault had a maverick personality, and his views frequently clashed with Boyington, who eventually left the AVG and rejoined the Marine Corps. He was recommissioned as a major and sent to New Caledonia in the South Pacific, where he mastered one of the service's newest planes, the bent-wing F4U Corsair.

A New Unit

While on convalescence following a leg injury, Boyington learned that the Marines badly needed to form new Corsair squadrons, and he organized an ad hoc unit comprising pilots and Corsairs dispersed by other units. The pilots' levels of experience ranged from combat veterans with several victories under their belts to new replacement pilots from the United States. Many of the pilots that Boyington pooled to form VMF-214 were known as misfits with reputations for discipline problems. The group quickly earned the nickname "the Black Sheep." Boyington once famously quipped, "Just name a hero and I'll prove he's a bum." A discipline problem himself, Boyington understood these pilots and trained them using tactics he'd learned as a Flying Tiger. It worked—in 84 days the Black Sheep destroyed or damaged 197 enemy planes.

His men called their leader "Gramps" because, though only in his early thirties, Boyington was ten years older than most of them. The press dubbed him "Pappy," and the name stuck as his reputation grew even larger. Between August and September 1943, Pappy had added 22 confirmed kills and was on his way to eclipsing Eddie Rickenbacker's World War I record of 26 downed planes.

On January 3, 1944, however, Boyington was shot down and believed to be dead. In fact, he was taken prisoner by the Japanese, who, knowing his identity, tortured him and refused to report his status to the International Committee of the Red Cross. His fate was not known until 18 months later, when Boyington emerged from a POW camp. Though the numbers are disputed by some historians, Boyington is officially credited with 28 kills, making him the leading Marine ace of the war. He was awarded the Medal of Honor.

Going to War in Style

Millions of Allied troops got to experience something they likely wouldn't have as ordinary civilians—an ocean voyage on a luxury passenger liner.

During World War II, scores of luxury passenger liners were pressed into service as troop-transport ships. Symbols of grandeur, privilege, and national prestige, these magnificent vessels were stripped of their finery, refitted, repainted, and recommissioned into decidedly more utilitarian craft that were indispensable to the Allied war effort.

The first large-scale troop transport involving luxury passenger ships left Halifax on December 10, 1939. Five camouflaged luxury liners—*Empress of Britain, Duchess of Bedford, Monarch of Bermuda, Aquitania, and Empress of Australia*—carried 7,400 soldiers of the Canadian Army 1st Division for deployment in Britain.

By war's end, millions of Allied soldiers had sailed aboard converted ocean liners to and from North America, Europe, Africa, Asia, and Australia. Here are looks at some of the more notable luxury ships that carried them to war in style.

Queen Mary

The jewel of the Cunard-White Star Line fleet, the *Queen Mary* stood alone among the classic ocean liners of its era, both as a peacetime luxury liner and as a wartime troopship.

It began its wartime service in May 1940. Painted gray while laid up in New York and fitted as a troopship in Sydney, the *Queen Mary* initially transported Australian troops to Scotland and Africa. In mid-1943, it was transferred to service in the Atlantic Ocean.

127

The *Queen Mary* was fast—at the outbreak of the war it was the undisputed holder of the shipping industry's Blue Riband award for the fastest trans-Atlantic liner. With a cruising speed of 28.5 knots, it often sailed unescorted, as German subs could not catch or keep up with it. The *Queen Mary* was also huge. It was eventually fitted for a capacity of 15,000 troops—equivalent to a whole division—and in July 1943, it established a record for the most people on a single voyage when it set sail with 16,683 soldiers and crew on board.

In all, the *Queen Mary* sailed 569,429 miles and carried a total of 765,429 military personnel during the war, surviving both a collision with a British light cruiser and a $250,000 bounty placed by Adolf Hitler. It returned to civilian service in 1946, but not before making 11 voyages bringing war brides to the United States and Canada.

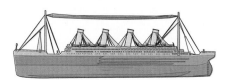

America (USS West Point)

The queen of the United States Line fleet, the *America* was launched on August 31, 1939, and entered into service in 1940. Within a year, the 33,560-ton liner was commissioned as an AP (armored personnel) ship by the U.S. Navy, renamed the USS *West Point*, and in only 11 days, refitted and put into service as the navy's largest troopship.

Nicknamed the "Gray Ghost" because of its wartime, dull gray makeover, the *West Point* could carry up to 8,175 troops at average speeds of approximately 25 knots. Through the course of the war, it would log distances equivalent to 16 circumnavigations of the globe while enduring several perilous encounters with the enemy. Off of Singapore in January 1942, Japanese bombers attacked the *West Point* and came within 50 yards of scoring a direct hit. The ship also faced air attacks in the Red Sea, off Australia, and at Port Suez. It also had close calls with submarines, including one that fired torpedoes across its bow off Brazil.

The *West Point* was decommissioned in early 1946 after having transported more than 350,000 troops to ports of call all around the world.

Manhattan (USS Wakefield)

Another ship of the United States Line fleet, the *Manhattan* was chartered by the U.S. government in 1940 to bring American citizens home from warring Europe. It was later commissioned by the U.S. Navy in May 1941 for use as a troop carrier and renamed the USS *Wakefield*.

Capable of carrying up to 6,000 passengers, the *Wakefield* often traveled as a "lone wolf," unescorted and fast enough to outrace German subs. Despite this, the *Wakefield* had a rough go of it early in its commission. It was traveling with the West Point off Singapore when alert Japanese bombers attacked the *Wakefield*, scoring a direct hit that resulted in five fatalities.

In September 1942, the *Wakefield* caught fire in the North Atlantic and was towed while still burning to Halifax, where it was almost capsized by a torrential storm.

The *Wakefield* was eventually repaired and recommissioned in February 1944 to continue serving as a troop transport. It finished out the war sailing primarily in the Atlantic Ocean, carrying 217,237 troops and passengers before being decom-missioned in June 1946. Unlike the *Queen Mary* and the America, however, the *Wakefield* never returned to its former glory as a luxury liner. It remained in the navy's possession, in reserve and out of commission, until released in 1959 and scrapped five years later.

A Wing and a Prayer

Thousands of World War II's aviators spent their days in small wooden boxes, didn't use parachutes, and couldn't fire a gun—yet managed to save hundreds of lives.

By 1939, the era of the homing pigeon was mostly history. Radio technology had improved over the Great War era, when war pigeons were in their heyday, making the winged messengers a thing of the past. But for men on a hot battlefield or partisans operating behind enemy lines, sometimes getting a message back to headquarters wasn't so simple. On such occasions, a humble carrier pigeon was a unit's best friend.

Allied Birds

Great Britain used carrier pigeons, also known as homing pigeons, more than any other military. An estimated quarter million of His Majesty's feathered subjects flew in service from 1939 to 1945.

The U.K.'s Royal Air Force (RAF), Civil Defense, and Home Guard made good use of pigeons, and even the home front did its part to support England's littlest wingmen: Pigeon racing was prohibited, birds of prey were hunted along the English coasts, and pigeon corn was rationed.

Homing pigeons were employed in a variety of jobs. RAF bombers and reconnaissance aircraft were equipped with pigeons so that if a plane had made an emergency landing, the birds could alert headquarters and a rescue could be launched. They were also extremely useful for sending secret messages to London. In occupied France, for example, resistance fighters regularly sent homing pigeons from the French coast to London with messages describing the dispositions of German units stationed along potential Allied landing zones.

The U.S. Pigeon Service, a branch of the Signal Corps, assigned more than 3,100 officers and men to manage some 54,000 pigeons during World War II. About a dozen pigeon companies were activated, and most were deployed overseas in all theaters.

Axis Birds

Germany had used carrier pigeons for many years. According to MI5, Britain's secret service agency, SS Führer Heinrich Himmler, a pigeon aficionado, made a "pet" project of air-dropping carrier pigeons to German spies stationed in England. One declassified MI5 report noted that two clandestine German birds had been captured, commenting wryly, "Both birds are now prisoners of war working hard at breeding English pigeons."

High-Flying Heroes

One British pigeon, named Scotch-Lass, was dropped into the Netherlands with a secret agent and arrived, wounded, back in London bearing 38 microfilm images. Another, a hen named Mary, was wounded several times during her five-year career before being killed in action. The pigeon White Vision flew more than 60 miles across churning waters near Scotland to deliver a distress message from a downed PBY (patrol bomber) flying boat in October 1943.

One Italian campaign veteran, a cock named G.I. Joe, flew 20 miles from the town of Colvi Vecchia, which British troops had occupied ahead of schedule, to U.S. bomber headquarters. The bird arrived in time to halt a planned bombing run on the city that would have brought down friendly fire on the Brits. The 5th Army's commander, Lieutenant General Mark Clark, estimated that G.I. Joe saved the lives of as many as 1,000 men. The bird was awarded the U.K.'s highest honor for a war animal, the Dickin Medal of Gallantry, nicknamed the "Animal's Victoria Cross."

Of course, pigeons were susceptible to the kinds of hazards every bird faces. For example, pigeon "10601" of the Royal Canadian Air Force, deployed from Allied submarines, accomplished many missions but was eventually brought down by a bird of prey. To neutralize the German spy threat, MI5 actually set up a falcon program to catch eastbound pigeons in one high-threat area. But through winds, hostile fire, and even the occasional enemy barn owl, these feathered messengers made contributions to the war efforts on both sides.

The Hundred Years' War—in Five Minutes

Although the name would suggest otherwise, the Hundred Years' War was not actually one drawn-out, century-long fight. Rather, it was a period of time containing multiple episodes of conflict.

Beginning in 1337 and ending in 1453, the ongoing trouble between England and France persisted through 116 years with unsuccessful attempts at truces, treaties, and peace between battles. Only in hindsight did historians combine these events under the descriptive title of the Hundred Years' War.

More than 400 years before the start of the Hundred Years' War, the region of France now known as Normandy welcomed a handful of Scandinavians. A century and a half later, with offspring of those feisty Vikings in tow, the Normans seized control of England and made it their own. Although settled comfortably into England, the Normans would maintain control over regions of France; however, with nationals from each region intermingling across country lines, ownership of and control over various regions became hazy. The question of who could claim rights to the different areas soon led to fierce bickering. These battles between France and England revved up and eventually turned into the Hundred Years' War.

First Things First

The first phase of the Hundred Years' War, often called the Edwardian War (1337–60), was instigated by a saga comparable to the Shakespearean struggle between the Montagues and Capulets: The son of King Edward I of England married the daughter of King Philip IV of France, and their mixed-breed child was in line to claim the French throne. But nobles of the land would not hear of someone they considered a mutt running their country, so they required that the royal ancestral lineage be paternal. All the while, control of France's Gascony and Calais regions was in dispute, and this monarchal twist only added to the tension. Ultimately, the English were the victors. The nobles of France, in a tizzy over their questionable leadership in the crown, couldn't unify their vast resources and relinquished control of huge swaths of land. The French became awash in national disenchantment.

Take Two

Although a peace treaty had been agreed upon, the two countries still couldn't get along. They were ready to fight again, so the second phase of the Hundred Years' War, called the Caroline War (1369–89), began. Spies aided France in learning England's fighting methods. Armed with this knowledge, the French soldiers were better mobilized with the hand-to-hand combat tools that had brought England victory in the first phase of the Hundred Years' War. This time, the win went to France. Once again, a truce was signed.

At It Again

The truce didn't last, and the Lancastrian War (1415–29) broke the peace as England attacked. Outnumbered, the English engaged their previous tactics and caught France off guard. The French succumbed to defeat. They refused to accept terms, however, so the truce was left up in the air. The Lancastrian War dragged on until Joan of Arc came forward to claim that she was destined to lead France to triumph over England. She did, in fact, win many a great victory, but her efforts didn't end the fighting, and she suffered execution at the hands of the English.

All's Well That Ends Well?

Although it had appeared England would be the final victor in this series of wars, at the last minute, Philip the Good of Burgundy pulled his troops from England and turned things around for France. Additionally, France finally one-upped England's weaponry advantage with its heavy artillery and mobile cannons. Although England had dominated 100 years earlier with longbows, the new technology made them seem crude by comparison. In the end, France earned back the majority of the land it had lost, with England retaining the city of Calais for roughly another century. During the final stretch of their century-long fight, their differences were settled with another, more lasting, treaty.

Ethan Allen

In the spring of 1775, when Ethan Allen and Benedict Arnold seized strategic Fort Ticonderoga at the start of the Revolutionary War, it's said that Allen claimed to have done so "in the name of the Great Jehovah and the Continental Congress."

That's the explanation Allen later alleged he gave British lieutenant Jocelyn Feltham, who challenged Allen's authority to enter Fort Ticonderoga. It sounds so red-white-and-blue: The homespun citizen-soldier cites God and Country as his authorities, and the rebellious redcoat backs off.

There is, alas, good reason to believe Allen's words were cruder. By all accounts, Allen was often crude—because he had to be. No mild-mannered dilettante could possibly have con-trolled Allen's Vermont militia, the Green Mountain Boys, who despised New York and the British alike. It is unlikely Allen would have invoked the Continental Congress, because he hardly respected its authority; it is far more likely he would have heaped scorn upon said Congress.

Another problem with the quote is its source—Ethan Allen himself. Lieutenant Feltham and one of Allen's junior officers agree, nearly to the word, that Allen actually yelled, "Come out of there, you damned old rat!"

How Old Is Old Ironsides?

It is the oldest warship in the U.S. Navy—and it's still in service. But the 44-gun frigate USS *Constitution*, the hero of the War of 1812, has survived only through numerous restoration efforts and a lot of patriotic passion.

The Legend's Service Record

Commissioned in 1797, this salty warrior made its name in an 1812 duel with Britain's HMS *Guerriere* off Nova Scotia. As *Guerriere* fought for its life, a U.S. sailor watched a British cannonball glance off *Constitution* and crowed, "Its sides are made of iron!"

The name ennobled a legend. A few months after *Guerriere* settled beneath the Atlantic waves, *Constitution* wrecked the speedy HMS *Java* off Brazil. Later, Old Ironsides would pummel two smaller vessels, HMS *Cyane* and HMS *Levant*, taking *Cyane* into U.S. service as a prize.

Old Ironsides actively served until 1855. The ship sat out the Civil War in New England after a quick escape from Annapolis. When the war ended, the Navy meant to tow it back, but Old Ironsides returned under its own power ten hours ahead of the steam tug. However, things were about to get ugly for the brave old frigate. By 1871, its sea legs were failing, so the Navy sent it to Philadelphia for repairs.

A Humiliated Hulk

Even given five years' lead time, no one managed to get *Constitution* shipshape for the 1876 Centennial. The job was completed a year late, with questionable workmanship and materials. On its last foreign cruise in 1879, Old Ironsides ran aground off Dover, England, then endured its worst indignity to date: It had to be hauled to safety by a British tug. The Navy sent it to Portsmouth, New Hampshire, and built barnlike bar-racks on its deck to house new recruits, much like Noah's Ark.

National Change of Heart

By 1900, some felt it was time to use Old Ironsides for target practice, but Congress realized that a national treasure was going to waste. Work finally started in 1906, and the deeper the crews dug, the worse decay they found—especially in the original timbers. A national campaign raised one-fourth of the million dollars needed to gut and restore the ship, which took until 1930. It spent the next four years showing the flag from Puget Sound to Bar Harbor, visited and loved by millions.

Later Restorations

Old Ironsides's home port is Boston, and there it spends most of its time as a beloved monument to the days of a young Republic. It underwent major renovations in the mid-1950s, early 1970s, and early 1990s, and by now even its original cannons have been replaced. Though only the keel and some ribs remain of the proud frigate that watched HMS *Guerriere's* masts fall, Old Ironsides gets better care entering its third century of service than it ever has. After far too much abuse, it finally has the dignity it is due.

Constitution is used today mainly for tours and education, but it still has an active-duty crew.

U.S. Veterans of the Revolutionary War

The American Revolution was fought between Great Britain and its 13 colonies on the eastern shoreline of North America. It began in 1775, lasted through the Continental army's victory in 1781 at Yorktown, and officially ended with the Treaty of Paris in 1783. Most of the Colonial soldiers who fought in the war were poorly trained and ill equipped; their British counterparts, on the other hand, were professional soldiers, many with actual battle experience. Thus, the phrase "the world turned upside down" describes the thoughts of many in Europe when news spread of a Colonial victory.

The last American veteran of the Revolutionary War was Colonial soldier Daniel F. Bakeman, who died on April 5, 1869, at age 109. Bakeman was 21 years old at the end of hostilities in 1781. He was the last survivor of the approximately 217,000 men who fought for independence from Great Britain, and he lived to see three other major U.S. wars, including the American Civil War.

Doin' the Duck and Cover

Growing up as an American child during the 1950s and '60s with the Cold War and the threat of Soviet attack looming made for an interesting, and often surreal, childhood.

Only a Sunburn

If you can remember hiding under your school desk with your hands over your head, then you are most likely a "Cold War survivor." For those who grew up in America during the 1950s, this "Duck and Cover" drill would have been part of their education. Intended to teach kids what they should do in the event of a nuclear blast, the drill even had its own mascot, Bert, an astute turtle who happened to carry his shelter on his back. Newsreels featuring Bert the Turtle warned children that the explosion of a nuclear bomb could "knock you down hard." But, if you only managed to "duck and cover" under a table or desk, you would be safely protected from the "sunburn" caused by the explosion.

Head in the Sand

These films were created by the Federal Civil Defense Administration, which had been given the responsibility of educating and protecting American citizens in the event of nuclear war. In effect, they were responsible for much of the American Cold

War propaganda produced during this time. One of their newsreels assured Americans that they would fare much better than the Japanese had in Hiroshima.

After all, they reasoned, unlike the Japanese, Americans were being given the information they needed to survive: If they just remembered to "clean under their fingernails, and wash their hair thoroughly," Americans were sure to be spared the effects of radiation.

A Culture of Fear

Fear had enormous impact on those growing up during the '50s and '60s. Many mid-century suburban towns had been built, in part, with the fear of Soviet attack in mind. Since it was expected that cities would be obliterated from falling bombs, the government encouraged suburban development as a way to scatter the population. Did you share a room with your brothers or sisters? During the Cold War, large families were encouraged as a way to fight the virus of communism—what better way to fight Soviet domination than to arm the country with a bursting population of wholesome capitalist American children? Birthrates soared to the highest levels of the century, with 29 million "baby boomers" born during the '50s.

The Cold War paranoia escalated after October 4, 1957, the day the Soviets launched *Sputnik* I, the world's first satellite. Suddenly, the nuclear threat emanating from Russia felt tangible. U.S. leaders decided the nation's best weapon against it was the youth of America, and so money was poured into schools and universities. No longer eggheads, scientists and engineers suddenly became the new elite.

Americans' response to the very real danger of nuclear war was sometimes surreal. In 1959, *LIFE* magazine ran a story featuring a couple planning to spend their honeymoon in a bomb shelter. The young couple was pictured surrounded by tin can provisions, which they took with them into their concrete "hotel." Children played with Sputnik toys and dressed as rockets for Halloween. Teenagers sported the latest swimsuit fashion, the bikini, named for the Bikini Island H-Bomb test site because its daring style was equally "dangerous."

On the Other Side

Since the fall of the Soviet Union it has been possible to learn how these years were experienced behind the Iron Curtain. Many Soviet people did not fear impending nuclear war nearly as much as Americans did. For a child in the Soviet Union, *Sputnik's* flight was a celebration of scientific progress, and America was the source of pop songs—not total destruction.

The Toledo War

The city of Toledo, by all accounts, is a lovely place. And no doubt it is; how else to explain why in 1835 and '36 Ohio and Michigan waged war—sort of—for the privilege to call Toledo their own.

As America began expanding westward, Congress passed the Northwest Ordinance in 1787, creating the Northwest Territory and establishing the north-south border between the future states of Michigan, Indiana, and Ohio. Michigan's southern border was to be a straight east-west line running from the southernmost point of Lake Michigan to the shore of Lake Erie. But the seeds of one of America's kookiest armed conflicts had already been planted when the border was drawn on a map charted in 1755 by John Mitchell.

Borderline Fraud

The Michigan-Ohio border drawn on the Mitchell Map (regarded as the most accurate map of eastern North America at that time and used extensively to determine international and state borders) intersected Lake Erie just north of the Maumee River. This placed the river mouth—and site of the future port settlement of Toledo—in Ohio.

However, when Congress passed the Enabling Act of 1802, allowing Ohio to begin the process of becoming a state, the language was more ambiguous when it came to the state's borders.

The situation got sticky in that year, when a fur trapper alerted Ohio delegates drafting a state constitution that Lake Michigan was actually farther south than on the Mitchell Map. Fearful of losing territory, the delegates engaged in some creative border drawing, slightly angling the boundary with Michigan to ensure the Maumee River basin remained Ohio property. The U.S. Congress never formally acted on Ohio's claim either way. The Ohio-drawn boundary remained unchallenged as Ohio gained statehood in 1803.

In 1805, the Michigan Territory was created, and its border with Ohio was determined by the provisions of the Northwest Ordinance (which predated the Mitchell Map). By then, cartographers had figured out where Lake Michigan really was, and Michiganders, armed with updated maps, contested Ohio's claim by declaring ownership of the Maumee River basin.

Settlers moving into the area, unsure if they were now Michiganders or Ohioans, petitioned for a resolution to the competing claims but were sidelined with the coming of the War of 1812.

The Toledo Strip

In 1817, U.S. Surveyor General and former Ohio governor Edward Tiffin moved to settle the issue by ordering William Harris to survey the border as drawn in the Ohio constitution. Miffed at Tiffin's blatant partiality, Michigan Territory governor Lewis Cass commissioned John Fulton in 1819 to survey the line as defined by the Northwest Ordinance. The resulting Harris and Fulton Lines created a narrow stretch of no-man's-land five miles wide at the Indiana border and eight miles wide at Lake Erie. Dubbed the Toledo Strip, the 468-square-mile area contained decent farmland in its western half, but swampland mostly surrounded the backwater port settlement of Toledo. Not much worth feuding about, it may have seemed, but feud the two sides did. For the next 15 years, Michigan assumed de facto jurisdiction over the strip, including Toledo, while Ohio refused to cede its claim to the territory.

The dispute came to a head in 1835 during Michigan's bid for statehood—the final obstacle being the continued squabble over the Toledo strip. Michigan Territory governor Stevens Mason proposed negotiations to settle the issue in January. Uninterested, Ohio governor Robert Lucas established a county government in the strip. Enraged, Mason enacted legislation making it illegal for Ohioans to conduct government activity in the strip. Undaunted, Lucas dispatched 300 Ohio militia to claim the land. Unimpressed, Mason marched a force of 250 Michigan militia to stop them. Michigan and Ohio were suddenly at war.

It's War! (Sort Of)

For a week in April, the two armies slogged through the Maumee swamps unable to find each other. Instead of battling it out, the two sides settled for hurling profanities at each other from opposite banks of the Maumee. Then came the farcically named Battle of Phillips Corners on April 26, in which a band of Michigan militia set upon a group of camping Ohio surveyors. Ohio accused the Michiganders of unleashing a barrage of gunfire at the surveyors; Michigan claimed its soldiers had fired a few rounds skyward as the Ohioans scrambled to the woods. No one was injured in the skirmish, which either confirmed Michigan's version of events or exposed them as awful shots.

Warfare consisted primarily of sheriff 's posses from both sides bullying civil servants and local residents. In one incident in July, Michigan sheriff Joseph Wood arrested Ohioan Benjamin Stickney. Stickney's son Two (the elder Stickney preferred to number his sons rather than name them) attacked Wood and stabbed him in the thigh with a penknife, thus making Wood the war's lone casualty (he survived). The only real fighting of the war occurred between drunken Buckeyes and Wolverines in Toledo's brawl-infested saloons.

Who Really Won?

By summer's end, President Andrew Jackson grew weary of the backwoods brouhaha. After failed attempts to mediate a solution, Jackson chose a side to end the conflict—and with an election coming up, he sided with Ohio. Jackson sacked Mason in August and offered Michigan the western three-quarters of the Upper Peninsula in return for giving the Toledo Strip to Ohio. Michigan refused.

Jackson denied Michigan statehood until it ceded the strip. In December of the next year, facing bankruptcy and desperate for a share of federal cash earmarked for the states, Michigan finally signed the so-called Frostbitten Convention, accepting the federal government's offer and ending the Toledo War.

Having secured Toledo, which many thought would become a great gateway to the American West with the building of the

Erie Canal, Ohio was deemed the victor. But Michigan arguably fared better, growing wealthy from the bountiful timber, copper, and iron reserves in the Upper Peninsula territory that was once considered the booby prize for losing the war. Though no lives were lost, both sides gave up a little dignity in this farcical bureaucratic conflict.

Mutiny or Mistake?

Military casualties aren't always the result of combat. A commander's own men can turn against him, too, whether by accident or by design.

Being a military commander sometimes means becoming a target, whether by the enemy or their own troops. Here are some commanders who found themselves dead at the hands of the men serving directly beneath them.

Nadir Shah

Cause of death: mutiny

Persian leader Nadir Shah ruled his empire during the 1700s, leading his army to victory over the Mongols, Turks, and other rivals. His career came to a sudden stop when his military-appointed bodyguard murdered him. Since Shah was said to be cruel, the other troops weren't too torn up about his death.

Colonel David Marcus

Cause of death: mistake

In the late 1940s, U.S. Army Colonel David Marcus decided to join Israel's army, as the nation was in the midst of a tense war.

Marcus was helping build a road from Tel Aviv to Jerusalem. One night, he couldn't sleep. He got up to go for a walk and an Israeli Army guard opened fire and killed him. Apparently, the guard saw a bed sheet wrapped around Marcus and thought he was an Arab fighter. His story was made into the 1966 film, *Cast a Giant Shadow*, starring Kirk Douglas.

Colonel John Finnis

Cause of death: mutiny

This one may not be a shock: Colonel John Finnis was slaughtered while lecturing his troops about insubordination. In 1857, the British Indian Army commander went to address his men after learning they were moving toward mutiny. Unfortunately, his message didn't carry much weight—the troops killed him mid-speech.

General Thomas "Stonewall" Jackson

Cause of death: mistake

On May 2, 1863, famed Civil War Confederate General "Stonewall" Jackson went out with a small crew to scout out his team's upcoming path. When he came back to camp, a group of Confederate fighters thought he was a Yankee and shot him. He died eight days later.

Captain Pedro de Urzúa

Cause of death: mutiny

In the mid-1500s, Spanish Captain Pedro de Urzúa found himself on the wrong side of his troops while leading soldiers on a mission across the Andes. De Urzúa's men rallied together and decided to take their leader's life, pledging their allegiance instead to another uniting figure. The story was later told in the 1972 movie, *Aguirre, the Wrath of God*.

Captain Yevgeny Golikov

Cause of death: mutiny

In 1905, Captain Yevgeny Golikov died over a dispute about meat. His crew, apparently upset over the low quality of meat available onboard, grabbed Golikov and tossed him into the sea. The story was the inspiration for the 1925 silent film, *Battleship Potemkin*.

Maass-ively Brave
....

Today, the name Clara Maass is not well known. But at the turn of the twentieth century, practically everyone in the United States knew her as the nurse who risked her own life to help defeat the dreaded yellow fever epidemic.

Young Girl Grows Up

Clara Louise Maass was born in East Orange, New Jersey, on June 28, 1876. She was the daughter of German immigrants who quickly discovered upon their arrival in America that the streets were not paved with gold.

Clara began working while she was still in grammar school. At around age 16, she enrolled in nursing school at Newark German Hospital in Newark, New Jersey. Graduating from the rigorous course in 1895, she continued working hard; by 1898, she was the head nurse at Newark German.

Yellow Jack War

The Spanish-American War began on February 15, 1898. As wars go, the war was more period than paragraph, lasting just four short months. However, there was something deadlier to American troops than Spanish bullets: yellow fever. No one knew what caused yellow fever or how to stop it.

The Experiments

In April 1898, Maass applied to become a contract nurse during the Spanish-American War; in Santiago, Cuba, she saw her first cases of yellow fever. The next year she battled the disease in Manila, as it ravaged the American troops there. Yet when Havana was hit by a severe yellow fever epidemic in 1900, Maass once again answered a call for nurses to tend the sick.

In Havana, a team of doctors led by Walter Reed was trying to find the cause of yellow fever. They had reason to support a controversial theory that mosquitoes were the disease carrier, but they needed concrete proof. Maass volunteered for the tests, though no one is sure why she put herself up to it.

On August 14, 1901, after a previous mosquito bite had produced just a mild case of sickness, a willing Maass was bitten once again by an *Aedes aegypti* mosquito loaded with infectious blood. This time the vicious disease tore through her body. She wrote a feverish last letter home to her family: "You know I am the man of the family but pray for me..."

On August 24, Maass died. Her death ended the controversial practice of using humans as test subjects for experiments. But it also proved, beyond a doubt, that the *Aedes aegypti* mosquito was the disease carrier—the key to unlocking the sickness that scientists had been seeking for centuries. Yellow fever could finally be conquered, in part because of Maass's brave sacrifice.

CHAPTER 6

LAW AND ORDER

Beyond the Law

After World War II, the victorious Allies brought quite a few Nazi war criminals to justice. Through careful planning, dumb luck, or Allied oversight, quite a few also escaped due justice.

Milivoj Ašner, Croatian police chief: Affiliated with the brutal *Usta e* (Croatian SS-equivalent), he organized the deportation of many Serbs, Roma, and Jews to concentration camps. He avoided capture and became an Austrian citizen after the war, then moved back to Croatia, where life was comfortable until a researcher found him. Ašner bailed for Austria, which refused to extradite him for trial in Croatia. He died of dementia in 2011.

Alois Brunner, Austrian SS Hauptsturmführer: Assistant to the infamous Adolf Eichmann, he was responsible for deporting well over 100,000 people to their deaths. His name resembled that of captured (and executed) fellow war criminal Anton Brunner, making the search less intense. Alois escaped to Syria under a fake name. His status is unknown but he is presumed dead.

Anton Burger, Austrian SS Sturmbannführer: First a bureaucrat coordinating the deportation of French, Belgian, Dutch, and Greek Jews to concentration camps, he then became commandant of Theresienstadt concentration camp. He escaped first from an internment camp, then from postwar Austrian custody, and lived anonymously in West Germany until his 1991 death.

Mikhail Gorshkow, Russian-Estonian Gestapo interpreter: He stands accused of involvement in the murder of several thousand Jews in the Slutsk ghetto of Minsk, Belarus (then the U.S.S.R.). He evaded capture and emigrated to the U.S., becoming naturalized in 1963. After his history surfaced, the U.S. stripped him of citizenship for lying on his application. Gorshkow fled to Estonia, where a war crimes investigation proved inconclusive.

Dr. Aribert Heim, Austrian SS Hauptsturmführer: The 'Dr. Death' of Mauthausen concentration camp, he specialized in killing by injecting toxin into the heart, as well as dissecting live prisoners without anesthesia. Somehow overlooked by his American captors, he practiced medicine in West Germany until the heat was on, then fled to Egypt and converted to Islam. He died in Cairo in 1992.

Dr. Josef Mengele, German SS Hauptsturmführer (captain): One of the most infamous men of World War II, he practiced well-documented war crimes both in selecting victims for gas chambers, and in experimentation: live dissections of pregnant women, sewing twins together, and other monstrosities. Mengele fooled his American captors with a fake name, then escaped to Argentina. He looked over his shoulder until his 1979 drowning in São Paulo, Brazil.

Walter Rauff, German SS Sturmbannführer (colonel): The designer of the gas vans, trucks set up to pump the exhaust into the bed, killing the passengers (Jews and the disabled) with carbon monoxide poisoning. He devised this solution because simply shooting prisoners was doing emotional harm to the shooters. He snuck out of an American internment camp in Italy and escaped to Chile, where he died in 1984.

Fumbling Felons

Choose Your Designated Driver Carefully

Many people drink alcoholic beverages responsibly. If they do have a little too much, most designate a capable driver. In November 2007, however, a 41-year-old man in Clio, Michigan, made a few key mistakes in his choice. Despite the fact that the man selected his son as his designated driver, the two ended up stuck in the mud where police arrested them both. What happened? His son was only 13 years old and also drunk.

He Was Only Reporting a Robbery

A man in south Texas called police to report a theft in 2007. Nothing unusual about that, right? Yet, the man told police that two masked gunmen had kicked in his door and stole 150 pounds of marijuana. He then explained to police that he was wrapping the drugs for shipment when the gunmen arrived. When police investigated the "crime," they found 15 pounds of the stuff lying on the man's floor. Not only did police charge the man with felony possession of marijuana, but he also turned out to be an illegal immigrant.

Half-Baked

After robbing two convenience stores in an hour, a man went for a third caper just a few hours later on September 30, 2007, in Delaware. The thief used a note that read, "Give me your money I'll shoot you." It was the same note used in the previous robberies. This time, however, the 25-year-old robber left his demand behind, and it just so happened to be written on the pay stub from his job at a local bakery. Along with fingerprints, the stub included the thief's full name. Bail was quickly set at $31,000.

Funny Money

In Gary, Indiana, a cafeteria worker broke up a money counterfeiting scheme when she was handed a fake $20 bill. The culprit: a 10-year-old boy. He aroused her suspicion when he paid for lunch with the large bill, so she turned it in. Police say the boy enlisted the help of two of his friends and created the money on his home computer. In December 2005, the children faced charges of forgery and theft. The boys are the youngest counterfeiters the FBI has ever come across. Reportedly, the counterfeits were even pretty good.

A Condemned Man Leaves His Mark

Does the ghostly handprint of a coal miner convicted of and executed for murder still proclaim his innocence?

In 1877, Carbon County Prison inmate Alexander Campbell spent long, agonizing days awaiting sentencing. Campbell, a coal miner from northeastern Pennsylvania, had been charged with the murder of mine superintendent John P. Jones. Authorities believed that Campbell was part of the Molly Maguires labor group, a secret organization looking to even the score with mine owners. Although evidence shows that he was indeed part of the Mollies, and he admitted that he'd been present at the murder scene, Campbell professed his innocence and swore repeatedly that he was not the shooter.

The Sentence

Convicted largely on evidence collected by James McParlan, a Pinkerton detective hired by mine owners to infiltrate the underground labor union, Campbell was sentenced to hang. When the prisoner's day of reckoning arrived, he rubbed his hand on his sooty cell floor then slapped it on the wall proclaiming, "I am innocent, and let this be my testimony!" With that statement, Alexander Campbell was unceremoniously dragged from cell number 17 and committed, whether rightly or wrongly, to eternity.

The Hand of Fate

The Carbon County Prison of present-day is not too different from the torture chamber that it was back in Campbell's day. Although it is now a museum, the jail still imparts the horrors of man's inhumanity to man. Visitors move through its claustrophobically small cells and dank dungeon rooms with mouths agape. When they reach cell number 17, many visitors feel a cold chill rise up their spine, as they notice that Alexander Campbell's handprint is still there!

"There's no logical explanation for it," says James Starrs, a forensic scientist from George Washington University who investigated the mark. Starrs is not the first to scratch his head in disbelief. In 1930, a local sheriff aimed to rid the jail of its ominous mark. He had the wall torn down and replaced with a new one. But when he awoke the following morning and stepped into the cell, the handprint had reappeared on the newly constructed wall! Many years later Sheriff Charles Neast took his best shot at the wall, this time with green latex paint. The mark inexplicably returned. Was Campbell truly innocent as his ghostly handprint seems to suggest? No one can say with certainty, but the legend lives on.

Lizzie Borden Did What?

Despite the famous playground verse that leaves little doubt about her guilt, Lizzie Borden was never convicted of murdering her father and stepmother.

The sensational crime captured the public imagination of late-nineteenth century America. On the morning of August 4, 1892, in Fall River, Massachusetts, the bodies of Andrew Borden and his second wife, Abby, were found slaughtered in the home they shared with an Irish maid and Andrew's 32-year-old daughter, Lizzie. A second daughter, Emma, was away from home at the time.

Rumors and Rhyme

Although Lizzie was a devout, church-going Sunday school teacher, she was charged with the horrific murders and was immortalized in this popular rhyme: "Lizzie Borden took an ax and gave her mother 40 whacks. When she saw what she had done, she gave her father 41." In reality, her stepmother was struck 19 times, killed in an upstairs bedroom with the same ax that crushed her husband's skull while he slept on a couch downstairs. In that gruesome attack, his face took 11 blows. One cut his eye in two. Another that severed his nose.

Andrew was one of the wealthiest men in Fall River. By reputation, he was also one of the meanest. The prosecution alleged that Lizzie's motivation for the murders was financial: She had hoped to inherit her father's estate. Despite the large quantity of blood at the crime scene, the police were unable to find any blood-soaked clothing worn by Lizzie when she allegedly committed the crimes.

Ultimately Innocent

Lizzie's defense counsel successfully had their client's contradictory inquest testimony ruled inadmissible, along with all evidence relating to her earlier attempts to purchase poison from a local drugstore. On June 19, 1893, the jury in the case returned its verdict of not guilty.

Convicted Conspirators
in President Lincoln's Assassination

Most of us learned in school that on April 14, 1865, President Abraham Lincoln was shot by John Wilkes Booth. What you may not have known was that Booth did not act alone, and that the plot wasn't limited to killing Lincoln.

John Wilkes Booth, mastermind of Lincoln's assassination, was shot to death by Union soldier Boston Corbett while attempting to escape on April 26, 1865.

Lewis Powell stabbed U.S. Secretary of State William H. Seward, but Seward recovered. The assassinations of Seward and Vice President Johnson were part of the conspiracy, but these assassination attempts were unsuccessful. Convicted of conspiracy to commit murder and treason, Powell was hanged July 7, 1865.

George A. Atzerodt was assigned to kill Vice President Andrew Johnson, but he never got very close to the vice president; most historians believe he merely wimped out. His second thoughts didn't save him, and he was convicted of conspiracy to assassinate the president. Atzerodt was hanged July 7, 1865.

David E. Herold guided Lewis Powell to Seward's home. Convicted of conspiracy to commit murder and treason, Herold was hanged July 7, 1865.

Mary E. Surratt owned the boardinghouse where the conspirators met. Convicted of conspiracy to assassinate the president, Surratt was hanged July 7, 1865.

During early conspiracy plans, **Michael O'Laughlen**, a boyhood friend of Booth's, was assigned to help kidnap Lincoln. Convicted of conspiracy, O'Laughlin received a life sentence. He died of yellow fever in prison in 1867.

Edman Spangler held Booth's horse during the assassination. Charged with conspiracy to assassinate the president,

Spangler was sentenced to six years. Pardoned by President Andrew Johnson due to lack of evidence in March 1869, Spangler eventually died in 1875.

Dr. Samuel A. Mudd harbored Booth and Herold during their escape attempt. Mudd was charged with conspiracy and sentenced to life in prison, but he was pardoned by President Johnson in March 1869 for his lifesaving efforts at Fort Jefferson during a yellow fever outbreak in 1867. Mudd resumed his medical practice. He died of pneumonia in 1883.

Samuel Arnold was involved in the early plans to kidnap President Lincoln. He was convicted of conspiracy and sentenced to life in prison. President Johnson pardoned Arnold in March 1869 because of his minimal role and early attempt to break from the conspirators. Arnold died of tuberculosis in 1906.

John Surratt also participated in the early plans to kidnap President Lincoln. He remained a fugitive until November 27, 1866, when he was apprehended in Alexandria, Egypt. He was charged with conspiracy, but a deadlocked jury resulted in Surratt's release in 1868. Surratt died of pneumonia on April 21, 1916. He was the last living convicted conspirator in the assassination of President Abraham Lincoln.

Heads Up: The Study of Phrenology

Sure, someone may look like a nice enough guy, but a phrenologist might just diagnose the same fella as a potential axe murderer.

He Had the Gall

There are bumps in the road and bumps in life. Then there are the bumps on our heads. In the last half of the nineteenth century the bumps and lumps and shapes of the human skull became an area of scientific study known as *phrenology*.

Early in the century, an Austrian physicist named Franz Joseph Gall theorized that the shape of the head followed the shape of the brain. Moreover, he wrote, the skull's shape was determined by the development of the brain's various parts. He described 27 separate parts of the brain and attributed to each one specific personality traits.

Gall's phrenological theories reached the public at a time of widespread optimism in Europe and North America. New and startling inventions seemed to appear every week. No problem was insurmountable, no hope unattainable. And a belief in physical science prevailed.

By mid-century, Gall's theories had spread favorably throughout industrialized society. What was particularly attractive about phrenology was its value as both an indicator and predictor of psychological traits. If these traits could be identified— and phrenology presumably could do this—they could be re-engineered through "moral counseling" before they became entrenched as bad habits, which could result in socially unacceptable behavior. On the other hand, latent goodness, intellect, and rectitude could also be identified and nurtured.

As it grew in popularity, phrenology found its way into literature as diverse as the Brontë family's writings and those of Edgar Allen Poe. It also influenced the work of philosopher William James. Famed poet Walt Whitman was so proud of his phrenological chart that he published it five times. Thomas Edison was also a vocal supporter. "I never knew I had an inventive talent until phrenology told me so," he said. "I was a stranger to myself until then."

Criminal Minds

Early criminologists such as Cesare Lombroso and Èmile Durkheim (the latter considered to be the founder of the academic discipline of sociology) saw remarkable possibilities for phrenology's use in the study of criminal behavior. Indeed, according to one tale, the legendary Old West figure Bat Masterson invited a phrenologist to Dodge City to identify horse thieves and cattle rustlers. A lecture before an audience of gun-toting citizenry ended with the audience shooting out the lights and the lecturer hastily departing through the back exit. In 1847, Orson Fowler, a leading American phrenologist, con-ducted an analysis of a Massachusetts wool trader and found him "to go the whole figure or nothing," a man who would "often find (his) motives are not understood." Sure enough, years later Fowler was proven to be on the money. The man was noted slavery abolitionist John Brown, and he definitely went the "whole figure."

Bumpology Booms

By the turn of the century, the famous and not so famous were flocking to have their skulls analyzed. Phrenology had become a fad and, like all fads, it attracted a number of charlatans. By the 1920s, the science had degenerated into a parlor game. Disrepute and discredit followed, but not before new expressions slipped into the language. Among these: "lowbrow" and "highbrow" describe varying intellectual capacity, as well as the offhand remark, "You should have your head examined."

Bad MOVE in Philly

Looking more like a war zone than the City of Brotherly Love, Philadelphia, Pennsylvania, was ignited under a police-induced firestorm in 1985.

They called themselves MOVE, short for the word "movement." Formed around 1972 by Donald and John "Africa" Glassey, the radical organization was comprised predominantly of African Americans who believed that a back-to-nature approach was central to living a full life. MOVE preached vigorously against the ills of technology and strongly embraced the idea of a society without government or police. Not surprisingly, their actions drew the suspicion and ire of the Philadelphia police. Eventually, tensions between the two groups would climax in the police bombing of MOVE's headquarters in 1985. To this day the incident is still vigorously criticized.

Hostilities started in 1978, when MOVE members were living communally in a house owned by Donald Glassey. Philadelphia police were wary of the group's actions there, and they released a court order demanding that MOVE, well, move. The radical group refused to relocate, however, and the ensuing confrontation claimed the life of Officer James Ramp and also injured several people. Nine MOVE members were subsequently tried, convicted of third-degree murder, and sentenced to 30 years in prison for their part in the shooting.

The Situation Escalates

By 1985, the remaining MOVE members were living in a row house at 6221 Osage Avenue. But they weren't quiet about their new residence—group members were heard shouting obscenities over bullhorns during the early hours of the morning. MOVE was suspected of hoarding weapons, and from what the police could see, they had even built a wooden bunker on their roof. Additionally, every window and door of their house was barricaded with plywood. Their actions not only made the police nervous, but they also frightened the group's neighbors, who turned to city officials for help.

Devastating Destruction

Many Philadelphians will never forget May 13, 1985. On that morning, an organized force of police, firefighters, and city officials converged on the residence in an attempt to force MOVE members from their antisocial haven. In short order, a standoff ensued; MOVE exchanged gunfire with the police. Possibly fearing a repeat of the 1978 incident, Philadelphia police planned a proactive, though ultimately fateful, strategy. At 5:30 that evening, they maneuvered a police helicopter over the house and released a bomb containing C-4 explosive. Although police claimed the bomb was only intended to destroy the bunker, it did far more than that.

Within minutes, the house was engulfed in a firestorm so powerful it would leapfrog streets and spread to adjacent homes. The only survivors to come out of the MOVE house were Ramona Africa (most MOVE members had taken the surname Africa) and a 13-year-old boy. In all, within four hours 11 people were dead (including five children) and 61 residences were decimated.

The Aftermath

Almost immediately, public opinion turned against the police. Questions soon arose: Why had the police dropped a bomb when they knew innocent women and children were inside the residence? Why had the fire department neglected to put out the fire? And other than noisemaking and unruliness, what had the MOVE group done at the Osage residence to merit such an attack? To this day, these questions remain largely unanswered.

Ramona Africa was charged with conspiracy, riot, and multiple counts of assault and served seven years in prison. In 1996, she was awarded $500,000 in a civil suit against the city. When asked her opinion of the bombing during a 2003 interview, Africa was blunt: "If the government is saying that their solution to a neighborhood dispute is to bomb the neighborhood and burn it down, then there wouldn't be a single neighborhood standing."

Fumbling Felons

Paper or Plastic?

Disguises can be tricky things—sure, they're great if they work, but every little detail has to be considered for that to happen. Especially details like, say, breathing.

An Arkansas thief found this out when he broke into an electronics store. He had forgotten his disguise, so he grabbed the first thing he could find—an opaque plastic bag. But not only did the bag prevent him from seeing where he was going, it also didn't allow in any air. The robber spent several minutes stumbling and tripping through the store, then finally collapsed and crawled away.

However, not willing to throw in the towel just yet, the hardy crook was back a few minutes later with yet another plastic bag disguise. This time, though, he had cut two eyeholes into the bag, which presumably let in some air as well. Fortified by fresh air, the crook managed to grab thousands of dollars' worth of electronic equipment.

When the cops reviewed the surveillance footage, they found that in his haste, the thief had neglected to remove the nametag from his clothing—a security guard's uniform from the mall where the store was located. The cops quickly corralled the crook, and took him to a place where he was issued a number to go with his name.

Criminal Quickies

A police department in Ottawa, Canada, had to expel a cadet from its officer training school when they discovered he'd stolen a car to get to class on time.

In Benecia, California, two armed robbers stuck up a credit union only to discover it was one of many "cashless" credit unions in the state. It would've paid to do some research first.

When Long Beach, California, armed robber James Elliot's revolver misfired, he peered down the barrel to check out the problem. He didn't survive.

More Fumbling Felons

When Honesty Is the Worst Policy

A clerk at a New Zealand food store was describing to police the man who had just robbed the store at gunpoint. Since the clerk had said that the man wasn't wearing a mask, the cop asked him to describe whatever he remembered to a police sketch artist.

As the clerk worked with the artist, it was clear that he had an amazing eye for detail. He noted specific features of the robber's face—a remarkable feat, especially for someone who had been held at gunpoint.

At last the artist finished and handed the picture to the investigating officer. The officer did an immediate double take. The clerk had described himself! When confronted with the fact, the clerk confessed that it was indeed he who had robbed the store. When the cop asked him why he had so accurately described himself to the sketch artist, the clerk responded: "I was just being honest!"

Wedding Bell Blues

An Alabama female police officer had previously worked prostitution stings, and she had seen a lot of odd things in her time. So, she didn't think much of it when, while she was working undercover, a man dressed in a tuxedo pulled up in a car alongside her and propositioned her. The officer played along, and soon the man found himself under arrest. Then the cop discovered why her "john" was dressed so nice: It was his wedding day. He had gotten married just a few hours before and had dashed out from the reception to buy more beer. But once out he apparently decided that booze wasn't enough to quench his thirst.

It's a good bet that her husband's arrest warrant was the one "gift" the bride didn't expect to receive on her wedding day!

No Sale

Sometimes you have to know when to walk away. A man from South Carolina bought substandard cocaine from his dealer. But instead of just feeling burned, he indignantly stormed into a police station. Throwing the bag of drugs disdainfully onto an officer's desk, the man demanded that the police arrest the dealer who had sold him the mediocre coke.

Further Fumbling Felons

A Developing Crime

Two boys from Louisville, Kentucky, stole a woman's Polaroid camera as she strolled through the park. Alerted by the woman's screams, a police officer gave chase, but the two thieves already had a head start.

Fortunately for the cop, the two boys had stopped and were taking pictures of one another. But much to their chagrin, the pictures that emerged from the Polaroid were all black, which, as many people know, is simply how Polaroid pictures look before they develop. Muttering about broken cameras, the boys continued on their way, occasionally stopping to take a photo. Each time a picture emerged from the camera all black, the thieves discarded it. All the pursuing cop had to do was follow the trail of rapidly developing photographs to find the technology-challenged crooks.

Who Ya Gonna Call?

With all of the informational resources available today, such as the Yellow Pages and the Internet, it's surprising that some people still have trouble finding the right person to contact for a particular task.

This was certainly the case for an Arizona woman who decided that she just couldn't stand her husband anymore. But instead of taking the obvious road and asking for a divorce, she contacted a company called "Guns for Hire" that staged mock gunfights for Wild West theme parks and the like. The woman asked them if they could kill her husband for her.

Now, while advertising is supposedly a key to a successful business, it's unlikely that a hired killer would go about broadcasting his or her services. After all, it tends to make the whole anonymity thing a bit difficult. On the plus side, however, at least the woman's prison sentence gave her a years-long vacation from her husband.

Crime Quickie: A mugger robbed a couple visiting a zoo in Blomfield, South Africa. Fleeing the scene, the mugger ran inside a tiger enclosure. The couple's belongings were recovered. The mugger was not.

LAW AND ORDER

Did You Know?

Prisoner Charles Justice was sentenced to death and executed at Ohio Penitentiary on November 9, 1911, courtesy of "Old Sparky," for the crimes of robbery and murder. Now here's the ironic part: Justice had been a prisoner in the same prison in 1900, where he helped clean the area where the electric chair was kept. Originally, the condemned prisoners would be bound to the chair by leather straps; if they strained under them and their skin broke contact with the chair's electrodes, the charge would jump the gap and severely burn their flesh. Justice made the helpful suggestion of using metal clamps to better secure prisoners. His ideas were put to use, and he was paroled for his efforts but ended up right back where he had been 11 years later.

True Tales of the Counterfeit House

On a hill overlooking the Ohio River in Monroe Township, Adams County, sat a house that wasn't what it seemed. Its modest size and quiet exterior hid countless architectural and historical secrets that earned it the nickname "The Counterfeit House."

In 1850, Oliver Ezra Tompkins and his sister, Ann E. Lovejoy, purchased 118 acres and built a rather peculiar house to suit the needs of their successful home-based business. Tompkins and Lovejoy were counterfeiters who specialized in making fake 50-cent coins and $500 bills. They needed

a house that could keep their secrets. Although passersby could see smoke escaping from the house's seven chimneys, only two of those chimneys were connected to working fireplaces; the others were fed by ductwork and filled with secret compartments. The front door featured a trick lock and a hidden slot for the exchange of money and products, and the gabled attic window housed a signal light.

The counterfeiting room was a windowless, doorless room in the rear of the house, accessible only through a series of trap-doors. A trapdoor in the floor led to a sizeable tunnel (big enough to fit a horse) that provided an escape route through the bedrock of the surrounding hills to a cliff. Although no records exist to support the imagined use of these features, local historians believe the reports to be true.

Visitors Not Welcome

While Lovejoy was in Cincinnati spending some of her counterfeit money, she was noticed by the police. A Pinkerton agent followed her home and watched as she opened the trick lock on the front door. He waited until she was inside, then followed her inside.

Immediately past the door, in a 10-foot by 45-foot hallway, Tompkins was waiting—and he beat the agent to death. To this day, bloodstains are still visible on the walls and floor. Tompkins and Lovejoy buried the agent's body in one of the nearby hills, and Tompkins used the hidden tunnel to escape to a friendly riverboat, collapsing the tunnel with explosives as he went. Lovejoy held a mock funeral for Tompkins and thus inherited his estate, although shortly after she went into debt and moved away.

Keeping Up the Counterfeit House

Although Tompkins never returned to the house, both his ghost and that of the agent were said to haunt it. Tourists reported seeing a man's shape in the front doorway and claimed that they felt unexplained cold spots and an unfamiliar "presence."

In 1896, a great-great uncle of Jo Lynn Spires, the last owner, purchased the property. It passed to Spires's grandparents in the 1930s, and Spires and her parents lived in the house with her grandfather. Although privately owned, the house was a tourist attraction, and Spires regularly kept the house clean, repaired, and ready for the stream of visitors that would trickle in each weekend. Unable to keep up with the repairs on the house, however, Spires moved into a trailer on the property in 1986. She continued to welcome approximately 1,000 tourists each summer.

In February 2008, windstorms caused severe damage to the house. One of the false chimneys blew apart, and the roof ripped off. Eventually, the house was demolished, and only its eerie memory lives on.

Whatever Happened to D. B. Cooper?

A parachute, a load of money, and a disappearing criminal combine in this strange tale.

On the day before Thanksgiving, 1971, in Portland, Oregon, a man in his mid-forties who called himself Dan Cooper (news reports would later misidentify him as "D. B.") boarded a Northwest Orient Airlines 727 that was bound for Seattle. Dressed in a suit and tie and carrying a briefcase, Cooper was calm and polite when he handed a note to a flight attendant. The note said that his briefcase contained a bomb; he was hijacking the plane. Cooper told the crew that upon landing in Seattle, he wanted four parachutes and $200,000 in $20 bills.

His demands were met, and Cooper released the other passengers. He ordered the pilots to fly to Mexico, but he gave specific instructions to keep the plane under 10,000 feet with the wing flaps at 15 degrees, restricting the aircraft's speed. That night, in a cold rainstorm somewhere over south-west Washington, Cooper donned the parachutes, and with the money packed in knapsacks that were tied to his body, he jumped from the 727's rear stairs.

Unanswered Questions

For several months afterward, the FBI conducted an extensive manhunt of the rugged forest terrain, but the agents were unable to find even a shred of evidence. In 1972, a copycat hijacker named Richard McCoy successfully jumped from a flight over Utah with $500,000 and was arrested days later. At first the FBI thought McCoy was Cooper, but he didn't match the description provided by the crew of Cooper's flight. Other suspects surfaced over the years, including a Florida antiques dealer with a shady past who confessed to his wife on his deathbed that he was Cooper—though he was later discredited by DNA testing.

Cooper hadn't hurt anybody, and he had no apparent political agenda. He became a folk hero of sorts—he was immortalized in books, in song, in television documentaries, and in a movie, *The Pursuit of D.B. Cooper*. In 1980, solid evidence surfaced:

An eight-year-old boy found $5,800 in rotting $20 bills along the Columbia River, and the serial numbers matched those on the cash that was given to Cooper. But while many leads have been investigated over the years, the case remains the only unsolved hijacking in U.S. history.

Even More Fumbling Felons

Printer Theft Leads to Sentence

Sometimes people aren't entitled to tech support on their electronics. Perhaps they didn't buy the warranty or register the equipment. Or in the case of Timothy Scott Short, maybe he stole the machine. Shortly after a printer that was used to make driver's licenses was stolen from the Missouri Department of Revenue in 2007, Short called the printer company's tech support wondering whether it was possible to buy a new part for it. A voice message and the suspect's phone number led police back to Short, who was charged with the felony of possessing document-making implements and theft.

No Dancing Matter

It all began on Andrew Singh's 2009 wedding day in Preston, Lancashire, England. A coach bus from Manchester was hired to transport three loads of wedding guests to the ceremony. On the way there, a car swerved into the bus causing a small collision. Oddly, the groom decided this was the perfect opportunity to come away with a bit of cash. He and his family sued the motor coach company, claiming that they had suffered injuries such as bruising and whiplash. But the case had no legs:

It was soon discovered that Andrew and his father were not actually passengers on the bus during the accident. A judge threw out their claim, and a police investigation was launched. The final straw was a video taken at the wedding reception, showing Andrew, his family, and festive wedding guests dancing, clapping, and cheering—and not looking very injured at all. The groom and his parents were convicted of conspiracy to defraud and perjury, and they were sentenced to a year in jail. Wisely, the bride ditched them all. Who's dancing now?

Con Man Quickies

In November 2006, a jail inmate in Austria climbed into a cardboard box and mailed himself to freedom. Two months later, a prison inmate in neighboring Germany did the same.

In 2008, Reginald Peterson called 911 to lodge a complaint. The charge? Apparently, a Jacksonville, Florida-based Subway sandwich shop employee left the sauce off his spicy Italian sub. In fact, he called 911 twice—first to complain about the lack of sauce and later to complain that police were slow in responding.

The Mad Bomber

Sure, he wanted revenge, but he also wanted to protect New Yorkers from their utility company.

Ninety miles North of New York City, George Metesky, an amiable-looking middle-aged man in a business suit, drove his car 80 feet from his driveway to the garage workshop at his family's house. He changed into coveralls and used gunpowder extracted from rifle bullets to craft what he called "units." He wanted New Yorkers to know that he had been wronged. When he meticulously packed away his tools at the end of the day, Metesky's bomb was ready.

The man who became the Mad Bomber nursed a grudge against his former employer, Consolidated Edison (Con Ed), New York's utility company. While working for Con Ed in 1931, Metesky suffered an accident and came to believe that he had been gassed and contracted tuberculosis as a result. Two things were indisputable: The illness left him unable to work, and Con Ed denied him workman's compensation.

A Little Attention, Please

More than 900 letters sent by Metesky to elected officials and newspapers failed to bring Con Ed to account. Frustrated, he devised an alternative plan. In November 1940, he left a pipe bomb outside a Con Ed plant on Manhattan's Upper West Side. A note read, "CON EDISON CROOKS, THIS IS FOR YOU." He signed it, "F.P." The bomb didn't go off, but Con Ed—and New York—had been warned.

The following September, an unexploded pipe bomb wrapped in a sock with a note signed "F.P." was discovered near Con Ed's headquarters. However, before Metesky could scare the city a third time, the nation entered World War II. New York City police received a letter from "F.P." outlining his patriotic claims: "I WILL Make no more BOmB UNITS for the Duration of the WAR . . . Later I WILl bring The con EDiSON to JUSTICE— THEy will pay for their dastaRdLy deeds."

New York saw no more bombs from "F.P." for nearly 10 years, although the threatening letters continued. Then, in March 1950, an intact bomb was found in Grand Central Station. "F.P." was back.

Clues

Metesky rapidly escalated his Con Ed war. A bomb blew up in the New York Public Library in April 1951, and another hit Grand Central. Between 1951 and 1956, Metesky placed at least 30 bombs. Although 15 people were injured by 22 bombs that exploded, no one was killed.

The lead detective turned to a criminal psychiatrist. Dr. James Brussel studied the case and concluded that the "Mad Bomber" was of Slavic descent, Catholic, and was burdened with an Oedipal complex. Detectives could find him outside the city living with a female relative. The NYPD was dubious. Dr. Brussel even told them that when they found "F.P.," he wouldn't come along until donning a buttoned double-breasted suit.

To trap "F.P.," the *New York Journal-American* encouraged him to submit his story. Metesky bit, and the story was printed. A Con Ed clerk had previously sifted through files of "troublesome" former employees and discovered Metesky. All this information added up to an identification. In January 1957, the cops drove to Waterbury, Connecticut, where Polish Catholic Metesky lived with his sisters. He opened the door in his pajamas and cheerfully admitted to being "F.P.," explaining that the initials stood for "Fair Play." Before he was arrested, he changed into a doubled-breasted suit.

Just What Is Insanity, Anyway?

Metesky grinned throughout his arraignment. He was sent to Bellevue Hospital for evaluation and ruled insane. He was committed to Matteawan State Hospital for the Criminally Insane without trial. On his release in 1973, Metesky told the *New York Times* that he wished he had stood trial. "I don't think I was insane," he said. "Sometimes . . . I wondered if there was something wrong with me, because of the extreme effort I was making." He reminded reporters that he was trying to help others. "If I caused enough trouble, they'd have to be careful about the way they treat other people." George "Fair Play" Metesky died in Waterbury in 1994 at age 90.

What Nellie Bly Found on Blackwell's Island

If you visit Roosevelt Island, you'll notice a building called the Octagon. These days, it's a posh condominium, but it was once the site of human injustice and chaos, nineteenth century style. Crackerjack reporter—and beloved New Yorker—Nellie Bly uncovered the story.

A little slip of land in the East River, Roosevelt Island was called Blackwell's Island during the eighteenth and nineteenth centuries. It was just farmland and hunting ground initially, but a prison was built in 1832, and several years later it was joined by the New York Lunatic Asylum, which was dominated by the Octagon Tower. The structure was beautiful, with an enormous spiral staircase and a domed, octagonal roof, but from the start, the asylum was grossly mismanaged. More than 1,700 mentally ill inmates were crammed inside (twice as many as should have been there), and although nurses were on duty, inmates from the nearby prison handled most of the supervision.

Over the next few decades, more prisons, asylums, and workhouses were built on Blackwell's, inspiring the island's new nickname: Welfare Island. Mortality was high because the care was so poor. Infants born there rarely lived to see adolescence. Any time spent on Blackwell's Island was too long for most.

The Girl's Got Sass

Help was on the way. Born in Pennsylvania in 1864, Elizabeth Jane Cochrane was a spitfire from the start. As a teen, she wrote an angry editorial to the *Pittsburgh Dispatch* about an article she found insulting to women. The editor was so impressed he hired her. Elizabeth assumed the pen name "Nellie Bly" (after a popular song) and lobbied hard for juicy stories. Although she landed a few, newspaperwomen at that time were relegated to the fashion and arts beats, a fate Bly fought against. Yearning for more substantive work, she left the Dispatch for New York City in 1887. She had bigger fish to fry.

Bly got a job at Joseph Pulitzer's *New York World* in hopes of significant stories. She already had one to pitch: She would feign insanity and get into the Women's Lunatic Asylum on Blackwell's Island. Everyone had heard about the conditions there, but no one had dared check it out. Bly's editors were duly impressed and gave their new employee the green light.

That night, Nellie checked into a Manhattan boardinghouse and commenced to freak everyone out. She acted bizarrely, dirtied her face, and feigned amnesia. Before long, the police came and took her away—straight into the heart of Blackwell's insane asylum.

From Bad to Worse

What the 23-year-old reporter found when she got there was worse than she had feared. For the next ten days, she endured the terrors and neglect that long-term inmates knew all too well. Life in the asylum was reduced to the animal level. Rotten meat and thin broth, along with lumps of nearly inedible dough, were all inmates were given to eat. And to wash it all down? Unclean drinking water.

Everyone was dirty, surrounded by their own filth and excrement from the rats that had free reign over the place. Baths consisted of buckets of ice water poured over the inmate's head, and the residents passed their days on cold, hard benches in stultifying boredom.

Bly's editors rescued her after 10 days, and Nellie wrote her exposé, a series of articles that eventually became a book called *Ten Days in a Mad-House*. The story blew up in the faces of the tin gods who controlled the prison and the asylum. Physicians and staff members tried to do damage control, but it was no use. A grand jury investigation commenced, and before long, new standards—many of which were suggested by Nellie herself—were implemented in institutions statewide. Monies were allocated, and the asylum on Blackwell's received long overdue repair and rehabilitation.

As for the young reporter, she would never have to go back to the fashion pages again. Bly continued to seek out adventure and remained a respected investigative reporter until she retired in 1895.

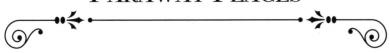

CHAPTER 7

FARAWAY PLACES

Chilly Accommodations

Want to spend some vacation time at an ice hotel? Better make reservations before spring comes, or you'll risk a melting room.

An ice hotel is a temporary hotel made entirely of ice and snow. Such hotels are open, logically enough, only during winter months; then they melt and are reconstructed the following year. Ice hotels are white-hot destination spots whose construction is based on some of the same theories used to build igloos: Blocks of ice and compacted snow from deep drifts are cut and arranged to build a structure that can protect and house the people using it. Two of the best-known hotels can be found in Sweden and Quebec.

ICEHOTEL, Sweden

ICEHOTEL in Sweden, located in the village of Jukkasjärvi, is 200 kilometers, or 125 miles, north of the Arctic Circle. It's the first ever hotel built entirely of ice. Because the hotel—or, hotels, one should say, as there is a new one every year—rests near the River Torne, there's an endless supply of clear water for construction each year.

The hotel uses 2,000 tons of ice in its construction. It covers more than 30,000 square feet. It is all ice, no kidding. The registration desk, tables, beds, chapel, art exhibition hall—all are made of ice.

So, what do you need to know before scheduling some sleep time in a hotel made of ice? First, you need to know that the temperature is kept between 15 and 24 degrees Fahrenheit. Concerned about keeping your extremities warm while spending an entire night in freezing temps? No worries—the hotel provides guests with thermal sleeping bags, and each ice bed is covered with reindeer skins. As long as you don't kick off the covers in your sleep, you should be able to stay free of frostbite for the night. A visit to the sauna in the morning is included in the price, along with a hot cup of lingonberry juice at your bedside.

Ice Hotel, Quebec

Inspired by the Swedish ICEHOTEL, designer Jacques Desbois built a similar hotel in 1996 in Sainte-Catherine-de-la-Jacques-Cartier, about 30 minutes outside of Quebec City. Quebec is a natural location for an ice hotel—the city hosts the annual Bonhomme Winter Carnival. The debut of the Ice Palace at the 1955 Winter Carnival was an early attempt at merging the pragmatic construction of igloos with the fantastical dreaminess of larger-scale ice buildings.

The Ice Hotel, Quebec, is usually made with 500 tons of ice and 15,000 tons of snow. It has 18-foot-high ceilings and theme suites, which are more numerous with each annual reconstruction. The Ice Hotel also has an Ice Bar sponsored by Absolut Vodka—a frosty delight. Guests of the Ice Hotel, Quebec, can expect ice beds similar to those at the ICEHOTEL, Sweden, but the Quebec version offers a thick foam mattress, as well as deer pelts, in addition to the requisite thermal sleeping bags. The Canadian version of the ice hotel houses two art galleries and a movie theater, which shows only "cool" movies.

Around the World

Sweden and Canada aren't the only countries to have ice hotels, of course. They can be built anywhere that's cold enough. Similar structures have been constructed in Norway, Romania, and Finland.

The Pyramids of Egypt

Magnificent they are, but they are not the oldest.

The Pyramids of Giza are the most famous monuments of ancient Egypt and the only structures remaining of the original Seven Wonders of the Ancient World. Originally about 480 feet high, they are also the largest stone structures constructed by humans. They are not, however, the oldest.

What's Older Than the Pyramids?

That glory goes to the prehistoric temples of Malta—a small island nation south of Sicily. The temples date from 4000 to 2500 BC. At approximately 6,000 years old, they are a thousand years older than the pyramids. Not much is known about the people who built these magnificent structures, but they were likely farmers who constructed the temples as public places of worship.

Because the Maltese temples were covered with soil from early times and not discovered until the nineteenth century, these megalithic structures have been well preserved. Extensive archaeological and restorative work was carried out in the early twentieth century by European and Maltese archaeologists to further ensure the temples' longevity. The major temple complexes are now designated as UNESCO World Heritage Sites.

Which pyramid is the oldest? That would be the Step Pyramid at Saqqara. It was built during the third dynasty of Egypt's Old Kingdom to protect the body of King Djoser, who died around 2649 BC. It was this architectural feat that propelled the construction of the gigantic stone pyramids of ancient Egypt on a rocky desert plateau close to the Nile. These pyramids, known as the Great Pyramids, were built around 2493 BC. The largest structure served as the tomb for Pharaoh Khufu.

Monumental Myths

Historians and conspiracy theorists have long debated the true meaning and origin of Stonehenge, the prehistoric series of stone monoliths located in England. Were they erected as an altar to aliens, a calendar for cosmic calculations, or a health spa?

No one really knows who built Stonehenge, because it was erected at a time before written language as we know it existed and word of mouth was, at best, unreliable. The ancient rumor mill claims that the Druids—a sect of Celtic priests—built the structures as a site for ceremonial sacrifices. This theory was posited by a couple of sixteenth century Stonehenge

antiquarians, John Aubrey and William Stukeley. But later archaeologists determined that the monuments predate the Druids by a thousand years, and it's also been noted that the sect worshipped in wooded areas, not stony enclaves.

Skyward Speculations

Because the entire structure has an out-of-this-world mystique, some imaginative analysts suggest it was built as a shrine to extraterrestrials, or that aliens themselves assembled the monuments. As evidence, these believers point to the fact that crop circles have repeatedly formed near the site. Still others are convinced that the monuments were created to act as a cosmic timepiece, and that the stones are precisely situated so the shadows they cast move like the hands of a clock.

Healing Among the Rocks?

Evidence of an ancient village on the outskirts of the site suggests the area was a place for the living, and that Stonehenge was a cemetery and memorial. Some researchers believe Stonehenge was a haven for wellness. The first stones moved to the site originally came from a bluestone quarry in west Wales that was used as a healing retreat. Archaeologists maintain that these stones were believed to have medicinal powers and were brought to Stonehenge for that purpose.

Byrd's First Antarctic Expedition

Richard Byrd introduced a new sort of Antarctic expedition.

U.S. Navy Rear Admiral Richard E. Byrd's 1928 Antarctic Expedition broke a rather long U.S. Antarctic exploration dry spell (88 years, since the last trip had been in 1840). Byrd was the first to integrate aerial photography, snowmobiles, and advanced radio communication into Antarctic study. With two ships, three airplanes, 83 assistants, lots of dogs, and a large heap of radio gear, Byrd set up a camp on the Ross Ice Shelf called Little America.

The expedition spent all of 1929 and January 1930 in Antarctica, making a flight over the South Pole and exploration Antarctic topography while recording a full year on the icy continent. Its tremendous success made Byrd famous.

The MVP of the Expedition

Norman Vaughan, Byrd's chief dog musher, dropped out of Harvard and became the first American to drive sled dogs in the Antarctic. Byrd gave him lasting credit by naming a 10,302-foot Antarctic mountain and a nearby glacier after him. (A few days before turning 89, Vaughan became Mount Vaughan's first ascender!)

At the 1932 Winter Olympics at Lake Placid, New York, Vaughan competed in dog mushing as a demonstration sport. Serving in the army in World War II, he became a colonel in charge of Greenland dogsled training and rescue. In one famed exploit, he rescued 26 downed airmen, then returned alone to recover the Norden bombsight, a top-secret device that was able to pinpoint targets using infrared radiation.

Born in 1905, Vaughan died on December 23, 2005, at age 100. He was the expedition's last living member.

You Live Where?

Ever hear of Boring, Maryland? How about Nimrod, Minnesota, or Boogertown, North Carolina? Many of the small towns that dot the United States have interesting stories (true or not) behind the oddball names. Here are a few stops to put on your next cross-country road trip.

- **Peculiar, Missouri**—As the story goes, 30 miles south of Kansas City was a small community needing a name. The folk put off naming their town—they didn't want to name it until their post office actually required it. The postmaster wrote the U.S. government requesting the regal-sounding name "Excelsior." Unfortunately, the name was already taken. The postmaster wrote time and time again for permission, using different names each time. Finally, in his exasperation he told them, "We'll take any name you have available as long as it's peculiar." Apparently, it stuck!

- **Wide Awake, Colorado**—One night when a group of miners were sitting around a campfire, they were trying to come up with a good name for their new settlement. After passing a bottle around late into the night, someone finally said, "Let's just turn in and talk about it more when we're wide awake." "That's it!" shouted one of the miners. "Let's call it Wide Awake!"

- **Toad Suck, Arkansas**—Before the Army Corps of Engineers completed a highway bridge over the Arkansas River in 1973, the most reliable way over the river was by barge. Next to the river stood an old tavern where many of the bargemen would pull over to drink rum and moonshine. As one version of the story has it, it was at this tavern that they would "suck on bottles until they swelled up like toads."

- **Accident, Maryland**—The town of Accident traces its history to 1750 when a local named George Deakins accepted 600 acres from King George II of England in relief of a debt. Deakins sent out two independent surveying parties to find the best 600 acres in the county—neither of which was aware of the other. By coincidence, they both surveyed the same plot, beginning at the same tree. Confident that no one else owned the property, Deakins named the tract the "Accident Tract."

- **Hell, Michigan**—There are several competing stories as to how Hell got its name. One story suggests that two traveling Germans stepped out of a stagecoach and remarked, "*So schönund hell!*" which loosely translates to "So beautiful and bright!" Hearing this, the neighbors focused on the latter part of the statement. Another story is that one of the early settlers, George Reeves, was asked what they should call the town. Ever the eloquent gentleman, Reeves replied, "For all I care, you can name it Hell!"

- **Ding Dong, Texas**—Despite evidence to the contrary, the town of Ding Dong was not named because it's located in Bell County. Nor was it named after Peter Hansborough Bell, the third Governor of Texas, nor for the Hostess snack cake. Back in the 1930s, Zulis and Bert Bell owned a country store, and they hired a creative sign painter named C. C. Hoover to put up a new sign. Hoover suggested that he dress up the sign by painting two bells on it with the words, "Ding Dong." The surrounding community quickly took to the name.

- **Tightwad, Missouri**—During the town's early days, a local store owner cheated a customer (who just happened to be a postman) by charging him an extra 50 cents for a watermelon. To get back at the proprietor, the postman started delivering mail to the newly dubbed town of Tightwad, Missouri.

Other Oddball Town Names:

Hot Coffee, Missouri

Truth or Consequences, New Mexico

Embarrass, Wisconsin *and* Minnesota

Knockenstiff, Ohio

Top-Secret Locations You Can Visit

There are plenty of stories of secret government facilities hidden in plain sight. Places where all sorts of strange tests take place, far away from the general public. Many of the North American top-secret government places have been (at least partially) declassified, allowing the average person to visit. Others can be visited only in the imagination.

Titan Missile Silo

Just a little south of Tucson, Arizona, lies the Sonoran Desert, a barren, desolate area where nothing seems to be happening. That's exactly why, during the Cold War, the U.S. government hid an underground Titan Missile silo there.

Inside the missile silo, one of dozens that once littered the area, a Titan 2 Missile could be armed and launched in just under 90 seconds. Until it was finally abandoned in the 1990s, the government manned the silo 24 hours a day, with every member being trained to "turn the key" and launch the missile at a moment's notice. Today, the silo is open to the public as the Titan Missile Museum. Visitors can take a look at one of the few remaining Titan 2 missiles in existence, still sitting on the launchpad (relax, it's been disarmed). Folks with extra dough can also spend the night inside the silo and play the role of one of the crew members assigned to prepare to launch the missile at a moment's notice.

Wright-Patterson Air Force Base

If you believe that aliens crash-landed in Roswell, New Mexico, in the summer of 1947, then you need to make a trip out to Ohio's Wright-Patterson Air Force Base. That's because, according to legend, the UFO crash debris and possibly the aliens (both alive and dead) were shipped to the base as part of a government cover-up. Some say all that debris is still there, hidden away in an underground bunker beneath Hanger 18.

While most of the Air Force Base is off-limits to the general public, you can go on a portion of the base to visit the National Museum of the U.S. Air Force, filled with amazing artifacts tracing the history of flight. But don't bother to ask any of the museum personnel how to get to the mysterious Hanger 18—the official word is that the hanger does not exist.

Area 51

Located in the middle of the desert in southern Nevada lies possibly the world's best-known top-secret location: Area 51. If you've read a story about high-tech flying machines, chances are Area 51 was mentioned. That's because the government has spent years denying the base's existence, despite satellite photos showing otherwise. In fact, it was not until a lawsuit filed by government employees against the base that the government finally admitted the base did in fact exist.

If you want to find out what's going on inside Area 51, you're out of luck. While the dirt roads leading up to the base are technically public property, the base itself is very firmly not open for tours—if an unauthorized visitor so much as sets one toe over the boundary line, he or she is subject to arrest or worse. Let's just say that the sign stating the "use of deadly force is authorized" is not to be taken lightly.

Los Alamos National Laboratory

Until recently, the U.S. government refused to acknowledge the Los Alamos National Laboratory's existence. But in the early 1940s, the lab was created near Los Alamos, New Mexico, to develop the first nuclear weapons in what would become known as the Manhattan Project. Back then, the facility was so top secret it didn't even have a name. It was simply referred to as Site Y. No matter what it was called, the lab produced two nuclear bombs, nicknamed Little Boy and Fat Man—bombs that would be dropped on Hiroshima and Nagasaki, effectively ending World War II. Today, tours of portions of the facility can be arranged through the Lab's Public Affairs Department.

Fort Knox

It is the stuff that legends are made of: A mythical building filled with over 4,700 tons of gold, stacked up and piled high to the ceiling. But this is no fairytale—the gold really does exist, and it resides inside Fort Knox.

Since 1937, the U.S. Department of the Treasury's Bullion Depository has been storing the gold inside Fort Knox on a massive military campus that stretches across three counties in north-central Kentucky. Parts of the campus are open for tours, including the General George Patton Museum. But don't think you're going to catch a glimpse of that shiny stuff—visitors are not permitted to go through the gate or enter the building.

Nevada Test Site

If you've ever seen one of those old black-and-white educational films of nuclear bombs being tested, chances are it was filmed at the Nevada Test Site, often referred to as the Most Bombed Place in the World.

Located about an hour north of Las Vegas, the Nevada Test Site was created in 1951 as a secret place for the government to conduct nuclear experiments and tests in an outdoor laboratory that is actually larger than Rhode Island. Out there, scientists blew everything up from mannequins to entire buildings. Those curious to take a peek inside the facility can sign up for a daylong tour. Of course, before they let you set foot on the base, visitors must submit to a background check and sign paperwork promising not to attempt to photograph, videotape, or take soil samples from the site.

Haunted Leap

They call it Ireland's most haunted castle, which says much in a country where the authorities must route new roads around fairy trees and mounds because they cannot find workers willing to disturb them.

A Violent History

Leap (rhymes with 'step') Castle is located in southern Co. Offaly, in the middle of the Republic of Ireland. It dates back to 1250. Legend says that the site was originally a druidic worship area. The Gaelic O'Carroll clan located their new stronghold with keen strategic eyes, commanding a pass leading to the province of Munster in southwestern Ireland. With a large tower and walls nine feet thick, besiegers would find Leap Castle a formidable target.

After 1532, the O'Carrolls entered a period of division, internal scheming, and backstabbing. In this weakened state, Leap was ripe for an English seizure. With a covert romance and a quick thrust of the sword—in the chapel during Mass, no less—the castle became Darby family property in 1659. They held it until 1922, when Irish rebels drove out the last Darbys with explosives and fire, leaving much of the castle in ruin.

Sordid Secrets Revealed . . . Somewhat

Around 1900, the Darbys hired workmen to clean out a windowless oubliette (dungeon pit). It contained hundreds of skeletons, evidently centuries old. Evidence indicated that prisoners were pushed through a hole into the oubliette, where they might die of immediate impalement on vertical spikes; those with less luck would miss the spikes and succumb to thirst in a lightless hole atop a pile of bones and rotting bodies. No one was sure whose bones they were, but Leap's past lords had clearly committed great cruelty and murder at the castle. Workers also found a mid-1800s pocket watch among the bones, raising suspicion that the oubliette had seen use in living memory. A nearby field called Hangman's Acre turned up iron hooks that suggested hangings by crueler means than the rope.

Then-châtelaine Mildred Darby was a devotee of the early 1900s occultism fad, and wrote of encounters with an inhuman elemental that looked and smelled like a decomposing corpse. Ghost hunters visiting the castle have reported similar sights and smells. Those who have seen the oubliette describe it as radiating profound evil, even with all remains removed.

In Recent Years

Leap lay vacant for some 50 years, boarded up and mostly avoided, though passersby sometimes reported seeing lights in the top windows. An Australian purchased the property in the 1970s, and brought a *bruja blanca* (white witch) from Mexico to cleanse all vileness from the castle. After getting plenty of exorcise, so to speak, the witch explained that the remaining spirits had no more ill intent, and wished to remain.

In the 1990s, the Ryan family purchased the property and started renovations. Soon after these began, Mr. Ryan broke his kneecap in one freak accident. A year later, after his recovery, the ladder he was working from somehow tilted away from the wall—forcing him to jump, which broke his ankle.

Not easily cowed by spirits and happenstances, the Ryans have continued restoration while living in Leap Castle, christening their baby daughter in the old chapel where such great malice was done long ago.

Today, visits to Leap are by prior appointment only, respecting that parts are inaccessible and that all is private property. A courteous phone call might obtain you an appointment. We recommend inquiring in nearby Roscrea—at the Heritage Centre in season, or if you prefer to do it Irish style, over pints at a local pub.

Tim Cahill, Adrenaline Junkie

Some travel the world and write about it. Some dine on haute cuisine; others lunch on sago beetle grubs. Meet Tim Cahill.

The Genre

Adventure travel is supposed to be interesting, but in recent years, Tim Cahill has taken it to the extreme. The Wisconsin native's travel and writing style has made his books, with titles like *Pecked to Death by Ducks* and *A Wolverine Is Eating My Leg*, simultaneously entertaining and weird.

The Guy

Chops help: Cahill swam competitively in college and is an advanced climber, caver, kayaker, diver, rider, and backcountry survivor. Guts help: Cahill will try just about anything.

One might expect an ego the size of Madagascar, but Cahill's writing doesn't reflect this. He doesn't get out of everything unscathed, and some situations in which he finds himself are just uncomfortable, embarrassing, or terrifying.

The Deeds

Over the course of a long career, Cahill's exploits have included:

• With Canadian endurance driver Garry Sowerby, setting a new speed record for driving from Ushuaia (Argentina's southernmost town) to Deadhorse (Alaska's northernmost road-reachable town).

• Doing yoga in Negril, Jamaica, while doing his all to avoid enlightenment.

• Teaching American grade school kids gorilla etiquette (which Cahill learned by doing, naturally).

• Learning the art of Neolithic spear throwing in east Idaho with a soft-spoken Mormon ex-Marine.

• Stalking the platypus by night in southern Australia.

• Tagging along with Tanzanian tribespeople as they negotiated a solution to recent donkey thefts.

• Kayaking in Glacier Bay while building-sized icebergs 'calved' from the glaciers.

• Drinking manioc beer among the jungle-dwelling Aguaruna of Peru while investigating the death of an idealistic young American.

- Hanging from a cliff in a diaper over the rapids of the Yellowstone River.

- Chowing on iguana eggs and meat in Honduras, then herding cattle from kayaks.

- Hobnobbing with New Guinea tribes where the men wear no clothes, but would feel naked without their penis gourds.

- Sea kayaking on wave-lashed California rocks with the Tsunami Rangers.

- Counting mounds of feces left by Iraqi troops in Kuwaiti buildings.

- Watching old movies with Marquesans who found Paul Newman's kissing hilarious.

- Looking for grizzlies in the Montana mountains—and finding them.

- Drinking kava (a mild intoxicant) with Tongans, while learning about giant clams and Tongan ways.

- Spelunking in a New Mexico cave where it snows gypsum.

- Taking a rubber boat onto Antarctica in five-foot swells, watching penguins bicker and bodysurf.

- Discovering witchcraft and getting drunk on pisco sours in southern Chile.

- Riding with low-rider enthusiasts and PCP users in East San José.

• Paragliding in Idaho, with an unerring ability to land in cow manure during training.

• Getting arrested in England along with the Dangerous Sports Club for watching them bungee-jump off a 245-foot-high bridge.

• Lunching on baked turtle lung with Aboriginal Australians near the tip of the Cape York Peninsula, Queensland.

• Investigating Olive Ridley sea turtle egg poaching in southern Mexico.

• Flying into a hurricane's eye with the Air Force.

• Watching Chinese archaeologists grapple with the challenges of eating lutefisk.

And those are just the ones he wrote about in books!

In Search of Immortality

Meet Gene Savoy, the "real Indiana Jones," who set off to discover cities and a whole lot more.

Born in Bellingham, Washington on May 11, 1927, Douglas Eugene "Gene" Savoy had no formal training as an archaeologist. But that didn't stop him from heading deep into the jungles of Peru. Once dubbed "the real Indiana Jones" by *People* magazine, Savoy discovered more than 40 lost cities in his career, including Vilcabamba, the last refuge of the Incas from the Spanish conquistadors.

Like movie hero Indiana Jones, Savoy's expeditions weren't entirely driven by archaeology. Savoy had grander plans—including finding the legendary city of El Dorado, where it was rumored that one could delve into the "ancient roots of universal religion" and the fabled fountain of youth.

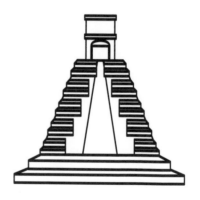

In 1969, Savoy left the jungle to search the sea. He captained a research ship and sailed around the world gathering information on sea routes used by ancient civilizations to prove that they could have been in contact with one another.

In 1984, Savoy returned to Peru, where he discovered Gran Vilaya, the largest pre-Columbian city in South America. On one of his last trips to Gran Vilaya, he unearthed a tablet with inscriptions alluding to King Solomon's ships that were sent to the biblical land of Ophir to gather gold for his temple. This tablet sent Savoy on what was perhaps his most ambitious adventure: to find the exact location of Ophir, find proof that the gold in Solomon's Jerusalem temple came from South America, and to learn the secret to immortality. Throughout his career, however, scholars scoffed at his theories and were skeptical of his findings.

But immortality eluded Savoy, and he died in 2007 in Reno, Nevada, where he was known as The Most Right Reverend Douglas Eugene Savoy, head of the International Community of Christ. Members of the church believed that staring at the sun would allow them to take in God's energy and become immortal (albeit damage their eyes)—a secret Savoy said was revealed to him in the jungles of Peru.

Say "Yes!" to
the World's Most Romantic Places to Propose

Proposals can pack a serious punch of romance. So, when it comes to popping the question, atmosphere is key. Below are some of the best spots the world has to offer.

The Pont Neuf (Paris, France): *Ay, me. Say amour.* Paris, France: the city of lights, the city of love, and most importantly the city of *ahhh*mazing proposals. An effervescent cheerleader for lovers everywhere, Paris is the epicenter for idyllic romance, so there is no surprise that its iconic structures follow suit. Pont Neuf, ironically meaning "new bridge," has now become the oldest bridge in France, making it a magical composition of old-world charm and new age love. Available at all hours and completely void of any expense to roam, this distinctive destination overlooks the Seine River and is most striking at sunset when the city takes on an amber glow.

Overlooking the Pitons (St. Lucia): Every guy wants his proposal to be perfect. After all, it will be the story his bride will retell over and over again to anyone who will listen—at least until it is topped by the actual big day. So, if the future groom wants to give his girl a gorgeous proposal, he needn't look any further than the gorgeous Gros and Petit Pitons in St. Lucia. Creating a picturesque backdrop of serenity, these volcanic wonders skyrocket over 3,000 feet from the admirably saturated blue waters below. Definitely a costlier alternative, this proposal spot can feature everything from a helicopter ride to a dip in the therapeutic

Hot Air Balloon Ride over the Masai Mara (Kenya, Africa): Hands down, one of the most well-known, well-loved adventures to experience, the calming nature of a hot-air balloon ride is undeniable, and when it's combined with the unmatched tranquility surrounding Africa's Masai Mara, there is no question of its magic. Sure, this kind of proposal can be rather expensive to accomplish, but when the outcome is a priceless result, it's impossible to compare.

A Sunset Sail in Bora Bora (Tahiti): Her timbers will surely shiver when you set sail at sunset in beautiful Bora Bora. The Le Meridien Bora Bora specializes in such proposals and can easily help any lovesick pup create a perfect memory. The epitome of Polynesian perfection, these sunset sails have all the ingredients for one outrageously quixotic "Will you marry me?" session: crystal clear waters, tropical breezes, sparkling champagne, and lovers at sunset.

Central Park (New York, New York): Central Park has been the mecca for romantic comedies the world over. Film and TV watchers really connect with New York–based love affairs, making Central Park a no-brainer for proposal settings. Unlike most seemingly romantic spots, Central Park doesn't have to be warm to be beautiful. It is just as magnificent—if not more so—when it's cold...just one more perk to this already incredible park. And for a true man with a plan (and a little extra dough), the Per Se restaurant in NYC is ideal. Overlooking the entire park, the splurge-worthy dinner for two needs to be booked at least two months in advance but provides a nine-course menu, as well as personal service, to create a one-of-a-kind environment.

Vertigo Restaurant (Bangkok, Thailand): One of Thailand's most appetizing treasures, the Vertigo Restaurant—part of the Banyan Tree Hotel—provides the utmost in exotic atmosphere, but its striking appeal is only the beginning. Where this killer of cuisine really dazzles is in its efforts to keep every proposal personal. The concierge will work with every Prince Charming to provide sentimental touches, such as her favorite flowers or top-choice wine. Set 61 floors above the bustling city of Bangkok, this delectable dining experience makes a statement all on its own, but once it is accompanied by a beau down on one knee, the results are nothing short of electric.

Geographic Oddities

The world's diverse landscapes, seascapes, and climates provide some interesting, and unusual, sightseeing opportunities.

Atacama Desert (Chile): This 600-mile stretch of coastal desert is so dry that if you die there, your corpse will barely decay. In some parts, no rainfall has been recorded by humans, but a million people still live in the region.

Bay of Fundy's Tides (Nova Scotia/New Brunswick, Canada): As the water moves in and out of this bay (which is the size of Connecticut), the difference between high and low tide reaches 52 feet. Think of a four-story building at water's edge, from which you could walk out on dry land at low tide. Come high tide, the entire building would be submerged.

Black Canyon of the Gunnison (Colorado): When a river flows across hard volcanic rock for millions of years, it cuts a deep course, and its sides don't collapse much. This explains the formation of the Black Canyon, which at one point is a quarter mile across and a half mile deep—from the air it looks like a vast, dark crack in the earth.

Blue Hole of Belize (Lighthouse Reef off Belize, Central America): Sixty miles out from Belize City, there's a circular reef in the shallow water about a quarter mile across, encasing a perfectly round, 400-foot-deep pool of midnight blue.

Cliffs of Moher (County Clare, Ireland): Here you can be rained on from above and below. The sheer cliffs rise more than 600 feet above sea level, and the surf's force is violent enough to send spray all the way to the top. When it's raining, one gets the stereo effect of being thoroughly drenched from both ends.

Dead Sea (Israel/Jordan): Called Yam Ha'maylach ("Salt Sea") in Hebrew, the Dead Sea is eight times saltier than seawater. You can't sink unless you're weighted down (not encouraged). It's referred to as "dead" because no fish or plant life can tolerate the high salinity, though its rich mineral content draws health enthusiasts from around the world.

Devil's Bath (North Island, New Zealand): Near Rotorua is a collection of geothermal attractions. The Champagne Pool (a hot, steamy, bubbly lake) spills over to create a bright, yellow-green pond called the Devil's Bath. It looks like a pool of molten sulfur and smells worse than one would expect.

Giant's Causeway (Antrim, Northern Ireland): Some 40,000 brown basalt columns (mostly but not all hexagonal) are packed together on and around a peninsula sticking out of Ireland. The result looks like stepping stones for a prehistoric titan and is one of the strangest and most popular sights in Ireland.

North Pacific Gyre Trash Vortex (Pacific Ocean): About a third of the way from California to Hawaii, a swirling ocean current collects garbage and doesn't easily let it go. This patch, now the size of Texas, consists mostly of floating plastic debris such as bottles and grocery bags. Unfortunately, as the pieces degrade, birds and fish eat them and die.

Punalu'u Black Sand Beach (Hawaii): Some places have beautiful beaches with sand that's white, or various shades of tan, or maybe even dark gray. Rarely, however, does one see sand that's as black as charcoal. The peculiar sand of Punalu'u is made of lava that exploded when it hit the water and has since been ground very fine. If that isn't strange enough, there are freshwater springs beneath the saltwater surf.

Uluru (Northern Territory, Australia): Formerly known as Ayers Rock, this enormous rusty sandstone monolith sticks up more than 1,100 feet from the desert floor and is about two miles wide. The rock is spiritually sacred to the Anangu (Australian Aborigines).

Uyuni Salt Flats (southern Bolivia): Roughly 25 times the size of the Bonneville Salt Flats in the United States, this saline landscape covers more than 4,000 square miles of Bolivia at an altitude of 12,500 feet. Because of brine just below the surface, any crack in the salt soon repairs itself.

Who Wants to Be a Troglodyte?

Lots of creatures sleep in cozy caves or burrows. Here are a few cave-based hotels and inns that are sure to make the animals green with envy.

Beckham Creek Cave Haven, Parthenon, Arkansas

This combination cave and cabin was built on a 530-acre estate in the Ozark Mountains. It makes the most of its gorgeous surroundings with big windows on its outside-facing wall, a rock waterfall, and a stalactite-studded ceiling. Below ground, guests have plenty of elbow room, and can divide their time between the 2,000-square-foot great room, spacious kitchen, game room, and five bedrooms tucked into rocky crevices.

Les Hautes Roches, Rochecorbon, France

This hotel is comprised of an eighteenth century castle and the caves that were created when stone was quarried for the Loire Valley chateau. Over the years, the cave was used as a refuge from war, to house monks from a nearby abbey, to grow mushrooms, and to store wine before becoming the first cave hotel in France. It now offers guests 15 rooms (12 of which are underground). A cave bar still contains the original fireplace and bread oven. Many guests use the hotel as a home base for exploring nearby mushroom caves.

Yunak Evleri, Urgup (Cappadocia), Turkey

The fairy towers and underground spaces carved out of Cappadocia tuff (soft volcanic rock) were once used as hiding places from armies. Today, tourists use them to find seclusion. One of the many hotels built into the area's caves, the Yunak Evleri contains 30 rooms. The hotel actually consists of six fifth- and sixth-century cave houses and a nineteenth century Greek mansion. Each room features whitewashed rock walls, hardwood floors, and antique Turkish rugs.

Kokopelli's Cave Bed & Breakfast, Farmington, New Mexico

This 1,650-square-foot, one-bedroom luxury cliff dwelling was not built in a natural cave. In 1980, geologist Bruce Black excavated the cave out of 40-million-year-old sandstone and turned it into an office 70 feet below the earth. During the next 15 years, Black worked to make the cave habitable. Bruce and his wife live in the cave, and since 1997, they've been welcoming paid guests. The cave entrance is located in the cliff face and is accessible via a steep path, 150 hand-hewn sandstone steps, and a short ladder. Guests are encouraged to pack light.

PJ's Underground B&B, White Cliffs, Australia

White Cliffs was home to the first commercial opal mine in Australia. Because it is self-supporting and easy to carve, the local sandstone made it easy for miners to create dugouts. In fact, most residents of White Cliffs dug their own dwellings with only a jackhammer and a wheelbarrow. Low humidity and constantly cool temperatures make the dugouts comfortable living spaces. PJs offers guests six underground rooms, an underground cottage, homegrown vegetables, homemade bread, and help arranging visits to the opal mines.

Las Casas Cueva, Galera, Spain

In ancient times, when a woman from the Andalusia region of Spain discovered she was pregnant, her first chore was to dig the baby's bedroom out of the limestone walls of her cave. Today, a tourist complex known as Las Casas Cueva de Galera, sits at the top of the hilly, prehistoric village of Galera. It is the biggest cave hotel in the region. Each cave includes a fireplace, kitchen, and Jacuzzi.

America's Haunted Hotels

Looking for ghosts can be a tiring experience, and sometimes, while on the road, even the most intrepid ghost hunter needs a good night's sleep. But if a peaceful night is what you're looking for, you may want to go elsewhere. In these places, no one rests in peace!

Admiral Fell Inn (Baltimore, Maryland)

The Admiral Fell Inn, located just steps away from the harbor on historic Fell's Point in Baltimore, was named for a shipping family who immigrated from England in the eighteenth century. With parts of the inn dating back to the 1700s, it's a charming place with stately rooms, an intimate pub, and wonderful service. It is also reportedly home to a number of spirits.

The ghosts at the Admiral Fell Inn include a young boy who died from cholera, a woman in white who haunts Room 218, and a man who died in Room 413. Staff members claim this room is always chilly and has strange, moving cold spots.

Blennerhassett Hotel (Parkersburg, West Virginia)

The Blennerhassett Hotel was designed and built in 1889 by William Chancellor, a prominent businessman. The hotel was a grand showplace and has been restored to its original condition in recent years. These renovations have reportedly stirred the ghosts who reside there into action.

There are several ghosts associated with the hotel, including a man in gray who has been seen walking around on the second floor and the infamous "Four O'Clock Knocker," who likes to pound on guest room doors at 4:00 a.m. There is also a ghost who likes to ride the elevators, often stopping on floors where the button has not been pushed. But the most famous resident spirit is that of hotel builder William Chancellor. Guests and employees have reported seeing clouds of cigar smoke in the hallways, wafting through doorways, and circling a portrait of Chancellor that hangs in the library.

Stanley Hotel (Estes Park, Colorado)

The Stanley Hotel has gained quite a reputation over the years, not only as a magnificent hotel with a breathtaking view but also as the haunted hotel that provided the inspiration for Stephen King's book *The Shining*.

In 1909, when Freelan Stanley opened his grand hotel, it immediately began to attract famous visitors from all over the country. Today, it attracts a number of ghostly guests as well.

The hotel's most notable ghost is Stanley himself; his apparition is often seen in the lobby and the billiard room. Legend has it that Stanley's wife, Flora, still entertains guests by playing her piano in the ballroom. Many people have reportedly seen the keys moving on the piano as music plays, but if they try to get close to it, the music stops. Another resident ghost seems especially fond of Room 407, where he turns lights on and off and rattles the occupants with inexplicable noises.

The Lodge (Cloudcroft, New Mexico)

Opened in the early 1900s, The Lodge has attracted famous visitors such as Pancho Villa, Judy Garland, and Clark Gable. And since 1901, every New Mexico governor has stayed in the spacious Governor's Suite.

The Lodge is reportedly haunted by the ghost of a beautiful young chambermaid named Rebecca, who was murdered by her lumberjack boyfriend when he caught her cheating on him. Her apparition has frequently been spotted in the hallways, and her playful, mischievous spirit has bedeviled guests in the rooms. She is now accepted as part of the hotel's history, and there is even a stained-glass window with her likeness prominently displayed at The Lodge.

The hotel's Red Dog Saloon is reputedly one of Rebecca's favorite spots. There she makes her presence known by turning lights on and off, causing alcohol to disappear, moving objects around, and playing music long after the tavern has closed.

Several bartenders claim to have seen the reflection of a pretty, red-haired woman in the bar's mirror, but when they turn around to talk to her, she disappears.

Hollywood Roosevelt Hotel (Hollywood, California)

Hollywood boasts a number of haunted hotels, but the Hollywood Roosevelt, which opened in 1927, is the most famous. Since 1984, when the hotel underwent renovations, the ghosts have been putting in frequent appearances.

One of the hotel's famous ghosts is Marilyn Monroe. Her image is sometimes seen in a mirror that hangs near the elevators. The specter of Montgomery Clift, who stayed at the hotel in 1952 while filming *From Here to Eternity*, has been seen pacing restlessly in the corridor outside of Room 928, where he appears to be rehearsing his lines and learning to play the bugle, which was required for his role in the movie.

Staff members and guests have frequently reported other unexplained phenomena, such as loud voices in empty rooms and hallways; lights being turned on in empty, locked rooms; a typewriter that began typing on its own in a dark, locked office; phantom guests that disappeared when approached; and beds that make and unmake themselves. Those Hollywood ghosts can be so demanding!

The Anasazi

Across the deserts and mesas of the region known as the Four Corners (where Arizona, New Mexico, Colorado, and Utah meet), backcountry hikers and motoring tourists can easily spot evidence of an ancient people. The towering stone structures at Chaco Culture National Historical Park and the cliff dwellings at Mesa Verde National Park tell the story of a culture that spread out across the arid Southwest during ancient times.

The Anasazi are believed to have lived in the region from about AD 1 through AD 1300 (the exact beginning of the culture is difficult to determine because there is no particular defining event). They created black-on-white pottery styles that distinguish regions within the culture, traded with neighboring cultures (including those to the south in Central America), and built ceremonial structures called kivas, which were used for religious or communal purposes.

Spanish conquistadors exploring the Southwest noted the abandoned cliff dwellings and ruined plazas, and archaeologists today still try to understand what might have caused the Anasazi to move from their homes and villages throughout the region. Over time, researchers have posed a number of theories, including the idea that the Anasazi were driven from their villages by hostile nomads, such as those from the Apache or Ute tribes. Others believe that the Anasazi fought among themselves, causing a drastic reduction in their populations.

Today, the prevalent hypothesis among scientists is that a long-term drought affected the area, destroying agricultural fields and forcing people to abandon their largest villages. Scientists and archaeologists have worked together to reconstruct the region's climate data and compare it with material that has been excavated. Based on their findings, many agree that some combination of environmental and cultural factors caused the dispersal of the Anasazi from the large-scale ruins seen throughout the landscape today.

Faraway Places

Their Journey

Although many writers—of fiction and nonfiction alike—romanticize the Anasazi as a people who mysteriously disappeared from the region, they did not actually disappear. Those living in large ancient villages and cultural centers did indeed disperse, but the people themselves did not simply disappear. Today, descendants of the Anasazi can be found living throughout New Mexico and Arizona. The Hopi tribe in northern Arizona, as well as those living in approximately 20 pueblos in New Mexico, are the modern-day descendants of the Anasazi.

Machu Picchu's Discovery

In 1911, Machu Picchu was rediscovered. The ancient "Lost City of the Incas" was rediscovered by Hiram Bingham III, an American professor and explorer. Bingham had become fascinated by the prospect of finding lost cities in South America while he was traveling through Peru following an academic conference in Santiago, Chile. He organized the 1911 Yale Peruvian Expedition to search for Machu Picchu, the lost city of the Incan empire. A guide named Melchor Arteaga, who lived in the nearby valley, led Bingham's expedition to the city, which is situated on a mountain ridge nearly 8,000 feet above sea level. Bingham incorrectly identified Machu Picchu as the Incan empire's capital, which was actually another city, named Vilcabamba. He returned three times, supported by the National Geographic Society. He came to believe that the city was a major religious site, but contemporary archaeologists have determined that it was most likely a summer estate for Incan royalty to retire to. Bingham was later elected to the United States Senate, where he served two terms. He died in 1956 and is buried at Arlington National Cemetery.

The Mystery of Easter Island

On Easter Sunday in 1722, a Dutch ship landed on a small island 2,300 miles from the coast of South America. Polynesian explorers had preceded them by a thousand years or more, and the Europeans found the descendants of those early visitors still living on the island. They also found a strange collection of almost 900 enormous stone heads, or moai, standing with their backs to the sea, gazing across the island with eyes hewn out of coral. The image of those faces haunts visitors to this day.

Ancestors at the End of the Land

Easter Island legend tells of the great Chief Hotu Matu'a, the Great Parent, striking out from Polynesia in a canoe, taking his family on a voyage across the trackless ocean in search of a new home. He made landfall on Te-Pito-te-Henua, the End of the Land, sometime between AD 400 and 700. Finding the island well-suited to habitation, his descendants spread out to cover much of the island, living off the natural bounty of the land and sea. With their survival assured, they built ahu—ceremonial sites featuring a large stone mound—and on them erected moai, which were representations of notable chieftains who led the island over the centuries. The moai weren't literal depictions of their ancestors, but rather embodied their spirit, or mana, and conferred blessings and protection on the islanders.

The construction of these moai was quite a project. A hereditary class of sculptors oversaw the main quarry, located near one of the volcanic mountains on the island. Groups of people would request a moai for their local ahu, and the sculptors would go to work, their efforts supported by gifts of food and other goods. Over time, they created 887 of the stone moai, averaging just over 13 feet tall and weighing around 14 tons, but ranging from one extreme of just under four feet tall to a behemoth that towered 71 feet. The moai were then transported across the island by a mechanism that still remains in doubt, but that may have involved rolling them on the trunks of palm trees felled for that purpose—a technique that was to have terrible repercussions for the islanders.

When Europeans first made landfall on Easter Island, they found an island full of standing moai. Fifty-two years later, James Cook reported that many of the statues had been toppled, and by the 1830s none were left standing. What's more, the statues hadn't just been knocked over; many of them had boulders placed at strategic locations, with the intention of decapitating the moai when they were pulled down. What happened?

A Culture on the Brink

It turns out the original Dutch explorers had encountered a culture on the rebound. At the time of their arrival, they found 2,000 or 3,000 living on the island, but some estimates put the population as high as 15,000 a century before. The story of the islanders' decline is one in which many authors find a cautionary tale: The people simply consumed natural resources to the point where their land could no longer support them. For a millennium, the islanders simply took what they needed: They fished, collected bird eggs, and chopped down trees to pursue their obsession with building moai. By the 1600s, life had changed: The last forests on the island disappeared, and the islanders' traditional foodstuffs disappeared from the archaeological record. Local tradition tells of a time of famine and even rumored cannibalism, and it is from this time that island history reveals the appearance of the spear. Tellingly, the Polynesian words for "wood" begin to take on a connotation of wealth, a meaning found nowhere else that shares the language. Perhaps worst of all, with their forests gone, the islanders had no material to make the canoes that would have allowed them to leave their island in search of resources. They were trapped, and they turned on one another.

The Europeans found a reduced society that had just emerged from this time of terror. The respite was short-lived, however. The arrival of the foreigners seems to have come at a critical moment in the history of Easter Island. Either coincidentally or spurred on by the strangers, a warrior class seized power across the island, and different groups vied for power. Villages were burned, their resources taken by the victors, and the defeated left to starve. The warfare also led to the toppling of an enemy's moai—whether to capture their mana or simply prevent it from being used against the opposing faction. In the end, none of the moai remained standing.

Downfall and Rebound

The troubles of Easter Island weren't limited to self-inflicted chaos. The arrival of the white man also introduced smallpox and syphilis; the islanders, with little natural immunity to the exotic diseases, fared no better than native populations elsewhere. As if that weren't enough, other ships arrived, collecting slaves for work in South America. The internal fighting and external pressure combined to reduce the number of native islanders to little more than a hundred by 1877—the last survivors of a people who once enjoyed a tropical paradise.

Easter Island, or Rapa Nui, was annexed by Chile in 1888. As of 2009, there are 4,781 people living on the island. There are projects underway to raise the fallen moai. As of today, approximately 50 have been returned to their former glory.

CHAPTER 8

YOU ARE WHAT YOU EAT

12 Items at a Feast of Henry VIII

Henry VIII, who ruled England from 1509 until his death in 1547, was known for his voracious appetite. Portraits of Henry show a man almost as wide as he was tall. When he wasn't marrying, divorcing, or beheading his wives (he was on his sixth marriage when he died at age 58), this medieval ruler dined like a glutton. He enjoyed banquets so much that he extended the kitchen of Hampton Court Palace to fill 55 rooms. The 200 members of the kitchen staff provided meals of up to 14 courses for the 600 people in the king's court. Here are some dishes served at a typical feast.

1. Spit-Roasted Meat

Spit-roasted meat—usually a pig or boar—was eaten at every meal. It was an expression of extreme wealth because only the rich could afford fresh meat year-round; only the very rich could afford to roast it, since this required much more fuel than boiling; and only the super wealthy could pay a "spit boy" to turn the spit all day. In a typical year, the royal kitchen served 1,240 oxen, 8,200 sheep, 2,330 deer, 760 calves, 1,870 pigs, and 53 wild boar. That's more than 14,000 large animals, meaning each member of the court was consuming about 23 animals every year.

2. Grilled Beavers' Tails

These tasty morsels were particularly popular on Fridays, when, according to Christian tradition, it was forbidden to eat meat. Rather conveniently, medieval people classified beavers as fish.

3. Whale Meat

Another popular dish for Fridays, whale meat was fairly common and cheap, due to the plentiful supply of whales in the North Sea, each of which could feed hundreds of people. It was typically served boiled or very well roasted.

4. Whole Roasted Peacock

This delicacy was served dressed in its own iridescent blue feathers (which were plucked, then replaced after the bird had been cooked), with its beak gilded in gold leaf.

5. Internal Organs

If you're squeamish, stop reading now. Medieval cooks didn't believe in wasting any part of an animal, and, in fact, internal organs were often regarded as delicacies. Beef lungs, spleen, and even udders were considered fit for a king and were usually preserved in brine or vinegar.

6. Black Pudding

Another popular dish—still served in parts of England—was black pudding. This sausage is made by filling a length of pig's intestine with the animal's boiled, congealed blood.

7. Boar's Head

A boar's head, garnished with bay and rosemary, served as the centerpiece of Christmas feasts. It certainly outdoes a floral display.

8. Roasted Swan

Roasted swan was another treat reserved for special occasions, largely because swans were regarded as too noble and dignified for everyday consumption. The bird was often presented to the table with a gold crown upon its head. To this day, English law stipulates that all mute swans are owned by the Crown and may not be eaten without permission from the Queen.

9. Vegetables

Perhaps the only type of food Henry and his court didn't consume to excess was vegetables, which were viewed as the food of the poor and made up less than 20 percent of the royal diet.

10. Marzipan

A paste made from ground almonds, sugar, and egg whites and flavored with cinnamon and pepper, marzipan was occasionally served at the end of a meal, although desserts weren't popular in England until the eighteenth century when incredibly elaborate sugar sculptures became popular among the aristocracy.

11. Spiced Fruitcake

The exception to the no dessert rule was during the Twelfth Night banquet on January 6, when a special spiced fruitcake containing a dried pea (or bean) was served. Whoever found the pea would be king or queen of the pea (or bean) and was treated as a guest of honor for the remainder of the evening.

12. Wine and Ale

All this food was washed down with enormous quantities of wine and ale. Historians estimate that 600,000 gallons of ale (enough to fill an Olympic-size swimming pool) and around 75,000 gallons of wine (enough to fill 1,500 bathtubs) were drunk every year at Hampton Court Palace.

Extra Cheddar, Please: Facts About Cheese

• Archaeological surveys show that cheese was being made from the milk of cows and goats in Mesopotamia before 6000 BC.

• Travelers from Asia are thought to have brought the art of cheese making to Europe, where the process was adapted and improved in monasteries.

• The Pilgrims had a supply of cheese onboard the *Mayflower* in 1620.

• The world's largest consumers of cheese include Greece (63 pounds per person each year), France (54 pounds), Iceland (53 pounds), Germany (48 pounds), Italy (44 pounds),

You Are What You Eat

the Netherlands (40 pounds), the United States (31 pounds), Australia (27 pounds), and Canada (26 pounds).

• The United States produces more than 25 percent of the world's supply of cheese, approximately 9 billion pounds per year.

• The only cheeses native to the United States are American, jack, brick, and colby. All other types are modeled after cheeses brought to the country by European settlers.

• Processed American cheese was developed in 1915 by J. L. Kraft (founder of Kraft Foods) as an alternative to the traditional cheeses that had a short shelf life.

• Pizza Hut uses about 300 million pounds of cheese per year.

• In 1886, the University of Wisconsin introduced one of the country's first cheese-making education programs. Today, you can take cheese-making courses through a variety of university agricultural programs, dairy farms, and cheese factories.

• Because they can produce large volumes of milk, butterfat, and protein, black-and-white (sometimes red-and-white) Holsteins are the most popular dairy cows in the U.S., making up 90 percent of the total herd.

• The Cheese Days celebration in Monroe has been held every other year since 1914. Highlights include a 400-pound wheel of Swiss cheese and the world's largest cheese fondue.

Beef Boner

Talk to the expert! Surviving employment in the meatpacking industry is no joking matter. When people realize how dangerous this job is, the laughter usually dies down.

Q: Most people would call you a butcher. In any event, you make a living cutting meat away from bones, and yet you still have all your fingers.

A: Many beef boners lose fingers—or worse. There is a reason many people who grow up around meatpacking plants are not interested in this line of work. I make a daily, concerted effort to be extremely attentive and careful. You have to watch for other people's knives as much as your own.

Q: Describe your boning knife.

A: It's similar to the steak knives you use at your dinner table. The blade is not very broad, it's smooth rather than serrated, and it has a very sharp tip. It's surprisingly small, considering how much animal flesh it slices up in a day. Some people imagine me waving a big meat cleaver all over the place, but that just happens in the movies.

Q: Do you cut up the whole cow?

A: Beef boners tend to specialize in one part of the animal—which is not usually a cow, but rather a steer. Whether I am working as a chuck boner,

YOU ARE WHAT YOU EAT

loin boner, ham boner, round boner, or blade boner, the basic work is the same: Cut out defects, bones, fat, and anything else people don't want to eat. You probably don't want me to go into too much detail here.

Q: What's the best way to avoid injury?

A: Stay alert and keep your boning knife sharp. It's true that a dull knife is more dangerous than a sharp one. Dull knives slip, and they make you work harder. The last thing I can be is fatigued.

An Indispensable Machine

Vending machines are a part of modern living, but they've been around much longer than you think.

Vending machines seem to be distinctly modern contraptions—steel automatons with complex inner workings that give up brightly packaged goods. The first modern versions were used in London in the 1880s to dispense postcards and books. A few years later, they were adopted in America by the Thomas Adams Gum Company for dispensing Tutti-Frutti-flavored gum on subway platforms in New York City. The idea of an automated sales force caught on quickly, and vending machines were soon found almost everywhere. The idea perhaps reached its peak in Philadelphia with the Automat in 1902. These "waiterless" restaurants allowed patrons to buy a wide variety of foods by plunking a few coins into a box.

Today, we think of vending machines as an everyday part of our lives. Americans drop more than $30 billion a year into them, and Japan has one vending machine for every 23 of its citizens. All kinds of products—from skin care items, pajamas, and umbrellas to movies, phones, and other digital devices—can be bought without ever interacting with a salesperson.

As high-tech as all that may be, the most remarkable thing about vending machines lies not in the modern era but in the distant past. A Greek mathematician and engineer named Hero of Alexander built the very first vending machine in 215 BC.! Patrons at a temple in Egypt would drop a coin into his device. Landing on one end of a lever, the heavy coin would tilt the lever upward and open a stopper that released a set quantity of holy water. When the coin slid off, the lever would return to its original position, shutting off the flow of water.

Would You Like Fried Worms With That?

Granted, to some people a Twinkie probably looks pretty weird. But at least Twinkies don't slither or smell like poo. Here's a sampling of some of the weirdest foods in the world.

Nutria

The nutria is a semi-aquatic rodent about the size of a cat with bright orange teeth. After World War II, they were sold in the United States as "Hoover Hogs." Since the animals chew up crops and cause erosion, in 2002 Louisiana officials offered $4 for every nutria killed. Still, their meat is rumored to be lean and tasty.

Uok

The coconut: Without it, the piña colada and macaroons wouldn't exist. Neither would the Uok, a golf ball-size, coconut-dwelling, bitter-tasting worm enjoyed by some Filipinos. Just pull one down from a mangrove tree, salt, and sauté!

Balut

If you're craving a midnight snack, skip the cheesecake and enjoy a boiled duck embryo. Folks in Cambodia will let eggs develop until the bird inside is close to hatching, and then they boil it and enjoy the egg with a cold beer.

Frog Smoothies

In Bolivia and Peru, Lake Titicaca frogs are harvested for a beverage affectionately referred to as "Peruvian Viagra." The frogs go into a blender with some spices and the resulting brown goo is served up in a tall glass. Turn on the Barry White . . .

Duck Blood Soup

Bright red goose blood is the main ingredient in this Vietnamese soup. A few veggies and spices round out the frothy meal.

A Wealth of Chocolate

In the sixteenth century, one of the treasures that Spanish conquistadors found was far more delicious than gold.

Chocolate Money

It's hard to believe now, with candy bars sold at nearly every store, but at one time, the cacao plant (used to make chocolate) was so valuable that the beans were used as currency. In the Americas, 100 cacao beans could buy a slave or a turkey hen; one bean could buy a tamale. The Spanish wrote that after the Aztecs conquered the tropical lowland areas of Central America in the 1400s, tribute was often paid in precious cacao beans. In fact, the Aztecs invaded the region of Xoconochco in part because of its production of high-quality cacao. Thereafter, local leaders had to pay tribute to the Aztec empire in precious items such as jaguar skins, the brilliant blue feathers of a cotinga bird, and hundreds of loads of cacao beans.

Not for Everyone

Aztec society was extremely stratified. Only the elite were allowed to drink xocoatl or "bitter water," a hot beverage made from cacao beans that had been ground into a paste and flavored with chilies, herbs, or honey. Franciscan missionary Fray Bernardino de Sahagún listed the different chocolate drinks served to the emperor: "green cacao-pods, honeyed chocolate, flowered chocolate, flavored with green vanilla, bright red chocolate, huitztecolli-flower chocolate, black chocolate, [and] white chocolate."

This hot chocolate was hardly the stuff of Swiss Miss. Once mixed, the chocolate was poured from container to container to produce the stiff head of foam that was an important element of the drink.

Modern Mexicans still enjoy chocolate in all its forms. In fact, a descendent of the ancient Aztec royal beverage remains popular in the form of Mexican hot chocolate, a frothy drink flavored with cinnamon, almonds, vanilla extract, and even chili powder.

The (Relatively) Harmless Truth
About Those Packets That Say "Do Not Eat"

It's better to add fruits and veggies to your diet than to take up a weird new culinary habit. But if you consider "do not eat" merely to be a friendly suggestion, you're in luck.

The stuff in those little packages is silica gel, which is a desiccant—a substance that absorbs and holds water vapor. Silica absorbs 40 percent of its weight in water and prevents moisture from ruining things.

Silica gel protects leather jackets from being damaged by moisture, prevents condensation from harming electronic equipment, and aids in retarding mold in foods such as pepperoni. The packets are especially useful during shipment, when a product starts in one climate (say, chilly Canada) and crosses several different locales before reaching its destination (say, balmy Florida).

But just how dangerous is it to eat? What would happen if you popped a silica packet into your mouth? The silica would instantly absorb as much of the moisture from your mouth as it could hold, which would make you very thirsty. If you were to swallow it, your throat would probably become parched, and then you would get a stomachache. It might also make your eyes and nasal cavity feel dry. But it wouldn't be deadly—

silica gel is nontoxic. In fact, the packets are more of a choking hazard than a toxin.

Now, if you decided to chow down on a bunch of silica packets, you would do some damage. But you probably couldn't afford all the pepperoni, leather jackets, and stereos it would take to make this a possibility.

A Tangy Tourist Attraction: The Mustard Museum

In a world of idiosyncratic people and places, Barry Levenson and the Mustard Museum just may be the spice of your life.

Triumph in the Face of Defeat

Their story begins on the early morning of October 28, 1986. The Boston Red Sox had just lost the World Series to the New York Mets, and just hours after the devastating undoing, despondent Red Sox fan Barry Levenson traipsed through an all-night grocery soul-searching for a clearer understanding of life. Crushed by his team's seventh game squander, Barry sought comfort in the condiment corridor. He waltzed by the pickles, the ketchups, the relishes, the horse radishes, and the mayos. When he hovered over the mustards, he heard a powerful voice: "If you collect us, they will come."

At the time of this zestfully zany epiphany, the Massachusetts native served as Assistant Attorney General for Wisconsin. Five years later, his mustard collection had grown so large that it warranted a bigger spotlight. So Levenson fully heeded that voice and devoted his attention full-time to the great golden hue, quitting law to start a museum with room enough to display the 1,000 jars he had already amassed.

Since then, tens of thousands of faithful minions and curiosity hounds have come to the Mount Horeb Mustard Museum, which opened April 6, 1992. Even after years of talking mustard, Levenson has not lost his interest in discussing the pungent paste's history, origins, varieties, and virtues, jawing at a pace approaching the speed of light.

The Wide World of Condiments

The Mount Horeb Mustard Museum represents the whole world of mustard powders and plants, from Azerbaijan to Zimbabwe, and quite a few places in between. Slovenian, South African, Italian, Scottish, Welsh, Russian, and Japanese mustards receive special attention, and Wisconsin mustards are spotlighted as well.

The museum houses more than 5,000 mustards and hundreds of items of mustard memorabilia, including mustard and hot dog art, literature, apparel, toys, coffee mugs, billboards, gift boxes, dispensers, model trucks, and souvenir buses and railroad cars. Mustard gift boxes are available to suit any occasion, and the tasting area allows visitors to sample local, regional, and exotic mustard sources.

Gourmet mustard patrons learn just how versatile, practical, and diverse mustard is—the museum teaches that there are more savory mustards than just the standard yellow variety. Thousands of hot pepper, garlic, herb, maple walnut, spicy apricot, black truffle, champagne, organic, dill, Dijon, and fruit variations await discovery.

Cutting the Mustard

About 35,000 visitors a year from across the globe come to the museum in search of insight. It has been featured on *The Oprah Winfrey Show*, HGTV's *The Good Life*, and the Food Network. Nevertheless, for Levenson, who has even authored a children's book entitled *Mustard on a Pickle*, no praise is too lavish for one of the world's most ancient spices and oldest known condiments, dating back to at least 3000 BC. when it was harvested in India. Today, Levenson says that about 700 million pounds of mustard are consumed worldwide each year, and that the U.S. uses more mustard than any other country.

But what if his beloved Red Sox hadn't lost the World Series to the Mets in October 1986? Would he have started collecting mustards that night—or ever? Would he have been so depressed as to wander the aisles of an all-night supermarket? Would the mustards sometime after have cried out so resonantly? The world may never know.

Munchie Mythology

Food: It's fodder for the best urban legends and old wives' tales. Myths range from complete nutritional nonsense to gluttonous celebrity gossip. Are you one of the gastronomically gullible?

You Are What You Eat

Bad Raps

MYTH: Chocolate causes acne.

TRUTH: No specific food has been scientifically proven to produce pimples—not chocolate, pizza, potato chips, or French fries. Acne's true cause is a buildup of dead skin cells within the pores. This can be triggered by hormones, environment, and heredity, but not by a Mr. Goodbar.

MYTH: Mayonnaise is the major cause of poisoning outbreaks from picnic foods.

TRUTH: Commercial mayonnaise is pasteurized and, thanks to ingredients like salt and lemon juice, has a high acid content that actually slows the growth of food-borne bacteria. Improperly handled meats and veggies in picnic salads and sandwiches are more likely to be your number one *Salmonella* suspects.

Brown Is Better Than White

MYTH: Brown eggs are more nutritious than white eggs.

TRUTH: Brown eggs come from hens with red earlobes, and white eggs come from hens with white earlobes. Crack through the outer shell, and brown eggs offer no better nutritive value, taste, or quality. Why are they more expensive? Brown eggs are usually a smidge larger in size.

MYTH: Brown sugar is healthier than white sugar.

TRUTH: Brown sugar is simply ordinary white table sugar that's turned brown by the addition of molasses. While molasses does contain certain minerals (calcium, potassium, iron, and magnesium), they are only present in negligible amounts. The real difference between brown sugar and white is only apparent in the taste and texture of your baked goods.

A Lie Your Parents Told

MYTH: Coffee will stunt your growth.

TRUTH: Research does not support the notion that drinking caffeinated coffee will hinder your height. That doesn't mean coffee belongs in a child's diet, however. Some actual adverse effects include bellyaches, nervousness, headaches, rapid heartbeat, and insomnia—all of which make for one crabby kid.

Into the Water

MYTH: You must wait an hour after eating before swimming.

TRUTH: Though swimming strenuously on a full stomach could lead to cramps, the chance of that happening to a recreational swimmer is quite small. One study of drownings in the United States found that less than 1 percent happened after the victim had recently eaten a meal. What you really need to avoid: eating or chewing gum while in the water. According to the American Red Cross, both activities can lead to choking.

Celebrity Stories

MYTH: Caesar salad was named for Julius Caesar.

TRUTH: The famous salad has no connection with that particular Caesar, or with Rome at all for that matter. Its creation is most often credited to Caesar Cardini, owner and chef of Caesar's Place in Tijuana, Mexico. His original recipe (concocted around 1924) contained romaine lettuce, garlic, croutons, Parmesan cheese, eggs, olive oil, and Worcestershire sauce. No anchovies!

You Are What You Eat

Hardy Little Tree

Averaging 26 to 49 feet in height, an olive tree is short and squat—about the size of an apple tree. Native to the Mediterranean, Africa, and Asia, this evergreen has silvery green, oblong leaves; a twisted, gnarled trunk; and beautiful clusters of white flowers.

The fruit of the olive tree is a small drupe. A drupe is a fruit in which a hard inner shell, or pit, is surrounded by outer flesh. Drupes are sometimes also called "stone fruits." Though olive trees do not bear fruit until they are 15 years old, they live for hundreds of years; in fact, an olive tree in Algarve, Portugal, has been radiocarbon dated at 2,000 years old.

The Israelites were not the only ancient people who valued the olive tree and its fruit. The Greeks in particular loved olives, and Homer included them in both the *Odyssey* and the *Iliad*. The Romans were also very fond of olives and integrated them into their diet (olives would eventually become a staple of Italian cooking). In Persia and across Arabia, olives were considered so healthy they were almost thought to be a miracle food.

So Many Uses

In ancient Israel, olive trees were grown inland because farmers believed coastal areas were not compatible to their growing, but this was later found to be untrue. In fact, olive trees seem to love poor soil. In rich soil, they are more likely to become diseased. Green olives were harvested from September to October in a process that involved shaking them off the trees and onto blankets to be sorted. Some olives were then pickled or packed in salt, but most were pressed for their oil.

Olive oil was consumed in all kinds of food, such as pottage (stew), bread, and salad. Its use went beyond the culinary, however. Olive oil was extremely important as a source of light. Every home in ancient Israel was lit with oil lamps, some quite primitive and some quite elaborate. The poor used clay lamp stands, while the rich owned impressive metal candelabras.

Olive oil was also used in grooming, both as a moisturizer for skin and a leave-in conditioner for hair. While the thought of oily hair may gross us out today, in ancient times it was considered quite attractive.

Mush-ruminations

• France was the first country to cultivate mushrooms, in the mid-seventeenth century. From there, the practice spread to England and made its way to the United States in the nineteenth century.

• In 1891, New Yorker William Falconer published *Mushrooms: How to Grow Them—A Practical Treatise on Mushroom Culture for Profit and Pleasure*, the first book on the subject.

• In North America alone, there are an estimated 10,000 species of mushrooms, only 250 of which are known to be edible.

YOU ARE WHAT YOU EAT

• A mushroom is a fungus (from the Greek word sphongos, meaning "sponge"). A fungus differs from a plant in that it has no chlorophyll, produces spores instead of seeds, and survives by feeding off other organic matter.

• Mushrooms are related to yeast, mold, and mildew, which are also members of the "fungus" class. There are approximately 1.5 million species of fungi, compared with 250,000 species of flowering plants.

• An expert in mushrooms and other fungi is called a mycologist—from the Greek word mykes, meaning "fungus." A mycophile is someone whose hobby is to hunt edible wild mushrooms.

• Ancient Egyptians believed mushrooms were the plant of immortality. Pharaohs decreed them a royal food and forbade commoners to even touch them.

• White agaricus (aka "button") mushrooms are by far the most popular, accounting for more than 90 percent of mushrooms bought in the United States each year.

• Brown agaricus mushrooms include cremini and portobellos, though they're really the same thing: Portobellos are just mature cremini.

• Cultivated mushrooms are agaricus mushrooms grown on farms. Exotics are any farmed mushroom other than agaricus (think shiitake, maitake, oyster). Wild mushrooms are harvested wherever they grow naturally—in forests, near riverbanks, even in your backyard.

• Many edible mushrooms have poisonous look-alikes in the wild. For example, the dangerous "yellow stainer" closely resembles the popular white agaricus mushroom.

• "Toadstool" is the term often used to refer to poisonous fungi.

• In the wild, mushroom spores are spread by wind. On mushroom farms, spores are collected in a laboratory and then used to inoculate grains to create "spawn," a mushroom farmer's equivalent of seeds.

• A mature mushroom will drop as many as 16 billion spores.

• Mushroom spores are so tiny that 2,500 arranged end-to-end would measure only an inch in length.

• Mushroom farmers plant the spawn in trays of pasteurized compost, a growing medium consisting of straw, corncobs, nitrogen supplements, and other organic matter.

• The process of cultivating mushrooms—from preparing the compost in which they grow to shipping the crop to markets—takes about four months.

• The small town of Kennett Square, Pennsylvania, calls itself the Mushroom Capital of the World—producing more than 51 percent of the nation's supply.

• September is National Mushroom Month.

• One serving of button mushrooms (about five) has only 20 calories and no fat. Mushrooms provide such key nutrients as B vitamins, copper, selenium, and potassium.

• Some experts say the taste of mushrooms belongs to a "fifth flavor"—beyond sweet, sour, salty, and bitter—known as umami, from the Japanese word meaning "delicious."

The Earl of Sandwich's Favorite Snack

The famed English statesman John Montagu named, but did not invent, the sandwich.

Legend holds that Montagu, the Fourth Earl of Sandwich, invented the tasty foodstuff that is his namesake. Montagu was a popular member of England's peerage in the eighteenth century, and it seems he had a knack for converting nouns into homage to his rank. The Hawaiian Islands were once known as the Sandwich Islands, thanks to explorer James Cook's admiration for the earl, who was the acting First Lord of Admirality at the time. And although it does seem likely that Montagu is responsible for dubbing the popular food item a "sandwich," he certainly was not the first to squash some grub between slices of bread.

A Sandwich by Any Other Name

It seems likely that sandwiches of one sort or another were eaten whenever and wherever bread was made. When utensils weren't available, bread was often used to scoop up other foods. Arabs stuffed pita bread with meats, and medieval European peasants lunched on bread and cheese while working in the fields. The first officially recorded sandwich inventor was Rabbi Hillel the Elder of the first century BC. The rabbi sandwiched chopped nuts, apples, spices, and wine between two pieces of matzoh, creating the popular Passover food known as charoset.

In medieval times, food piled on bread was the norm—prior to the fork, it was common to scoop meat and other food onto pieces of bread and spread it around with a knife. The leftover pieces of bread, called "trenchers," were often fed to pets when the meal was complete. Primary sources from the sixteenth and seventeenth centuries refer to handheld snacks as "bread and meat" or "bread and cheese." People often ate sandwiches; they simply didn't call them that.

It Is Named…Therefore It Is?

Regardless of the sandwichlike foods that were eaten prior to the eighteenth century, it appears that the Fourth Earl of Sandwich is responsible for the emergence of the sandwich as a distinct food category—but how this happened is unclear. The most popular story relates to Montagu's fondness for eating salted beef between pieces of toasted bread. Montagu was also known for his gambling habit and would apparently eat this proto-sandwich one-handed during his endless hours at a famous London gambling club. His comrades began to request "the same as Sandwich," and eventually the snack acquired its name.

The source that supports this story is *Tour to London*, a travel book that was popular at the time among the upper echelons of society. In one passage, the author of the book, Pierre Jean Grosley, claimed that in 1765, "a minister of state passed four and twenty hours at a public gaming-table, so absorpt in play that, during the whole time, he had no subsistence but a bit of beef between two slices of toasted bread. This new dish grew highly in vogue…it was called by the name of the minister who invented it." According to this scenario, "sandwich" initially referred to Montagu's preferred beef-and-bread meal and was subsequently used as an umbrella phrase for a variety of sandwich types.

YOU ARE WHAT YOU EAT

Hard Work and Hunger

John Montagu's biographer, N.A.M. Rodger, offers an alternate explanation for the rise of the sandwich. He argues that during the 1760s, when the sandwich was first called a sandwich, the earl was actually busy with government responsibilities and didn't have time to gamble much. He did, however, spend many nights working at his desk, during which time he liked to munch on beef and bread. It is possible, Rodger argues, that the sandwich came to be as a reference to the earl's tireless work ethic and general fondness for late-night snacking.

Sushi: The Hallmark of Japan's Fast Food Nation

While the combination of fish and rice has been a mainstay in Asian cuisine for millennia, sushi as we know it today began with an entrepreneurial mind, lots of street carts, and a little bit of sumo wrestling.

Complexities in Fish Fermentation

Modern sushi consumers are accustomed to sushi of the fast-food variety—throw rice on seaweed, add a strip of fish, and presto, you have a snack that is both nutritious and delicious. Yet the earliest type of sushi, known as *Narezushi*, took more than six months to prepare and was so smelly that it was eventually replaced by the stink-free sushi that we know today.

The sushi prototype actually originated in China and was then perfected in Japan. Before the days of refrigeration, innovations in fish preservation abounded. One popular method was to press fish between layers of salt for months at a time. At some point, it was discovered that fish would ferment faster if it was rolled in rice *and* pickled with salt. After the fermentation was complete, the rice was discarded, and the fish was eaten alone.

This method of fish preservation was popular in China and Southeast Asia, but it was only in Japan that the process eventually evolved into the snack-size morsels of sushi. The original Narezushi was created by a complex process that involved salting and pickling fish for more than a month, piling the pickled fish in between layers of cool rice, sealing everything into a barrel, then pickling some more. This drawn-out exercise created a sourness that was awful to smell but delectable to the taste buds.

It's difficult to popularize a food that's six months in the making, but this all changed with the invention of rice vinegar, which decreased fermentation time. Sushilike dishes were popular during Japan's Muromachi Period (1338–1573) and again during the Azuchi-Momoyama period (1574–1600). The first type of sushi to gain widespread popularity was the Oshizushi of Osaka, which was rectangular in shape and consisted of alternating layers of rice, fish, and sometimes pickled vegetables. Yet even this popular sushi, quite literally, stank.

Sushi Served Up Fast

The process of fermentation had to be abandoned completely before sushi could come out smelling like roses—or at least not like old fish. It wasn't until the rise of a large urban city, where fish could be caught and consumed in plenty within a 24-hour time frame, that modern sushi became possible.

In late eighteenth century Edo (Tokyo), it became common to place a strip of raw fish on a mound of rice. This is called *nigiri-zushi*, or handmade, sushi.

The popularization of nigiri is usually attributed to early nineteenth century entrepreneur and chef Yohei Hanaya. Legend has it that he was throwing a dinner party in 1824 when he realized he didn't have enough fish to 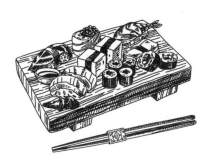 go around. As a solution, he placed small slabs of fish on large mounds of rice. The entrepreneur in Hanaya realized he'd just discovered a gold mine.

Hanaya transformed his off-the-cuff innovation into a fast-food phenomenon. He sold his nigiri sushi in carts throughout the streets of Edo, and soon others opened sushi carts of their own.

Hanaya opened his first cart in front of Ryukoku Temple, where frequent sumo wrestling tournaments created an ongoing glut of pedestrians. The idea was that sushi is a finger food that can quickly be prepared and eaten, right on the street. The fish in these early sushi were often cooked, marinated, or heavily salted, so dipping in soy sauce was not necessary. Sushi went from food carts to restaurants throughout Edo to restaurants throughout Japan and eventually gained its current status as a worldwide food favorite.

The word *sushi* actually refers to the pickled or vinegared rice, not the fish itself. *Sashimi* refers to the raw fish.

Deadly Nightshades

Do nightshade vegetables like tomatoes and eggplant contain a toxic acid? Or a toxic alkaloid?

The nightshade family goes far beyond tomatoes and eggplants, including potatoes, bell and chili peppers, and tobacco. There's also deadly nightshade, petunias, and the real-life mandrakes featured as shrieking nuisances in the Harry Potter series. Nightshades can get a bad rap, but is it deserved? That depends who you ask, although evidence suggests nightshades are fine.

Supermodel Gisele Bundchen and New England Patriots quarterback Tom Brady, a celebrity married couple who look very much like siblings, made news in 2016 with their radically healthy private chef diet. Their self-imposed limitations are extensive, including no fruit, nightshade vegetables, gluten, caffeine, "fungus," dairy, or MSG.

This long, mystifying list might register with some readers as bordering on orthorexia, an eating disorder in which the pursuit of a so-called healthy diet leads to obsessive tracking, avoiding, and phobic reactions to allegedly unhealthy foods.

But Bundchen and Brady have a dedicated private chef who offers up tailored meals to fit their purported needs, which definitely must make it easier not to choose foods that aren't part of your plan. Strangely, despite their laundry list of no-nos, Bundchen and Brady do still eat meat.

Their chef's shout-out to nightshades drew attention to this whole family of vegetables. Are they the new gluten, another food that is not harmful at all for about 99 percent of people?

You Are What You Eat

Should we take the word of someone who swaps a gluten-containing whole grain for trendy quinoa that actually doesn't have any more protein or less carbs than that comparable whole grain? Let's address some claims.

Nightshades cause migraines. Migraines are enigmatic and individual sufferers should choose for themselves when it comes to avoiding trigger foods. But there isn't a demonstrable link between nightshades and migraine. Other common triggers like chocolate or garlic have longer paper trails to back them up.

Nightshades contain a toxic alkaline. Certain specific nightshades, like potatoes that have turned green and need to be thrown away, do contain solanine. But almost all other vegetable nightshades are fine.

Nightshades contain a toxic acid. Nightshades have oxalic acid in small quantities, while the herbs parsley and chives contain much more of the acid by volume. Oxalic acid can stop your body from absorbing calcium but only if you consume very little calcium to begin with—and a *lot* of nightshades and other vegetables with oxalic acid.

Nightshades make arthritis worse. Arthritis is a frustrating condition that can lead sufferers to grasp for possible triggers, very similar to migraine. There's no evidence that nightshades worsen arthritis inflammation or pain. But again, individual sufferers can do their due diligence to find if they experience discomfort around any kind of food or substance.

Nightshades cause general inflammation. Inflammation has become a health buzzword without a clear definition. Acetaminophen is an anti-inflammatory drug; does that mean it cancels out a nightshade? In any case, there's no evidence that nightshades cause inflammation in people without an allergy to them. Chiles may make your mouth feel inflamed, but you can douse that with milk, unless you're Gisele.

Nightshades can upset your stomach. Certainly, you may find that specific nightshades, like starchy fried potatoes or hot chiles, upset your stomach. And everyone's digestive system is different. If you find that you feel sick or bloated after meals where you've eaten these foods, you can try avoiding them for a couple of weeks to see if you notice an improvement.

Nightshades aren't worth eating. The nightshade vegetables represent a huge portion of the human diet during the agricultural age. For a surprising number of Americans, the slice of tomato they get on a sandwich embodies one of their most consumed vegetable groups—after lettuce. This huge family of very nutritious vegetables should be part of the diet of everyone who doesn't have an explicit health reason to avoid it.

It's interesting to find the "unhealthy" potato among the eggplants, tomatoes, and peppers. Potatoes may be caloric, but they contain many nutrients, and they were even carried onto ships to help prevent scurvy because of their high levels of vitamin C. The alkaline substance that green potatoes produce is a natural defense against predators, which is what the spiciness of peppers and the other nightshade idiosyncrasies are probably for. It may be part of why this family of plants is so robust and durable.

The danger with drawing a circle around an entire family of foods and declaring them off limits for no reason, as many people have done with gluten as well, is that all humans must always eat every day. Placing restrictions on foods that are

You Are What You Eat

good for us only makes choosing foods feel even more frustrating overall, and that feeling of confusion and helplessness can create a bad feeling around eating. Food nourishes us and keeps us going, even nightshades.

In 2017, Tom Brady released a book that forms the cornerstone of his new pyramid scheme of wellness products. In between complete common-sense nonstarters like how people should eat more vegetables and get enough sleep, he throws in plugs for his Tom Brady brand of Tom Brady supplements. We can only hope one of them contains some nightshades to mix things up a little.

You Are What You Eat

Though not a chef, Anthelme Brillat-Savarin was the foremost food writer of his time. He touted cooking as a science, and his ideas about diet and obesity remain relevant to this day.

Had there been a Food Network in the late eighteenth and early nineteenth centuries, it's a safe bet Anthelme Brillat-Savarin would have hosted his own show. And that's saying something, considering he was not a chef at all, but rather a French attorney and politician.

Brillat-Savarin was born in Belley, a river town in Eastern France, in 1755. He followed his father into the law business, ventured into politics and in 1792 was elected mayor of his hometown. One of his deep beliefs was in the merit of capital punishment, a controversial cause about which he wrote and delivered speeches.

Brillat-Savarin's tour as mayor lasted just one year. Some in Paris saw him as a counter-revolutionary and seized his property. A bounty was placed on his head toward the end of the French Revolution, sending Brillat-Savarin to Switzerland, The Netherlands and eventually to the newly formed United States. He lived in Boston, Philadelphia, New York and Hartford, making his way by teaching French and violin lessons. He was actually playing first violin in New York's John Street Orchestra for a time.

A Foodie at Heart

For all his many talents, however, Brillat-Savarin was more inspired by food than perhaps anything else. His parents were both good cooks, and two of his brothers were gourmets. Brillat-Savarin was welcomed back to France in 1796 and soon gained a lifelong position in the Court of Cassation. While in Paris, Brillat-Savarin could focus on his writing, and that writing ventured well beyond the bounds of politics and law.

His most famous work held the long French title *Physiologie du Goût, ou Méditations de Gastronomie Transcendante; ouvrage théorique, historique et à l'ordre du jour, dédié aux Gastronomes parisiens, par un Professeur, membre de plusieurs sociétés littéraires et savantes*. It has been translated, and shortened, to *The Physiology of Taste* or *The Philosopher in the Kitchen*. It was published anonymously in December 1825, two months before Brillat-Savarin's death.

One of Brillat-Savarin's tenets was that food was, quite objectively, good or bad. That is, a person's reaction to a dish did not indicate anything at all about the quality of the dish. The dish was either of good, medium or poor quality to begin with. One's reaction to it told about the person, rather than the food itself.

And when it came to those dishes, Brillat-Savarin's belief that "you are what you eat" was in many ways a foreshadowing to what became known as the Atkins diet in modern times. He espoused a diet low in carbohydrates for weight loss and better health. He advised readers to stay clear of starches and flour-based dishes, claiming that they lead to greater rates of obesity. "All animals that live on farinaceous food grow fat willy-nilly," he wrote, "and man is no exception to the universal law."

Brillat-Savarin, who never married, died at age 70. He remains an inspiration to several food writers, TV personalities and critics of today. The dishes he loved were usually not lavish ones, but rather simple ones. His legacy also lives on in a cheese (Brillat-Savarin) and a cake (gateau Savarin) named in his honor.

Quotes from Brillat-Savarin

"Tell me what you eat, and I will tell you what you are."

"A dessert without cheese is like a beautiful woman with only one eye."

"A meal without wine is like a day without sunshine."

The Texas Pig Stand Introduces
Front Seat Dining

The drive-in restaurant wasn't born in California, nor did the McDonald brothers invent the fast food genre. Carhops, curb service, and the Pig Sandwich are what started it all.

In 1921, Texas mercantile wholesaler Jesse Granville Kirby made the proclamation, "People in their cars are so lazy that they don't want to get out of them to eat!" At the time, he was trying to hook Reuben Jackson, a Dallas physician, to invest $10,000 in a new idea for a roadside stand, one that paired the Lone Star State's love for the car with another pastime: eating barbecue.

For the era, Kirby's idea was revolutionary: Texans were to drive up to the food stand and make their requests for food directly from behind the wheel of their cars (or trucks, of course, this being Texas). A young lad would take the customers' orders directly through the window of the car and then deliver the food and beverages right back out to the curb. The novelty of this new format was that hurried diners could consume their meals while they were sitting in the front seat.

Convenience Over All

When Kirby and Jackson's Texas Pig Stand opened along the busy Dallas–Fort Worth Highway in the fall of 1921, legions of Texas motorists tipped their ten-gallon hats to what was advertised as "America's Motor Lunch." Prepared with tender slices of roast pork loin, pickle relish, and barbecue sauce, Kirby and Jackson's Pig Sandwich quickly gained a loyal following among cabbies, truckers, limousine drivers, police officers, and other mobile workers.

But curbside cuisine wasn't the only attraction at America's first drive-in restaurant. The daredevil car servers who worked the curb—or carhops, as someone coined the phrase—were a sight. "Carhops were very competitive," recalled Richard Hailey, successor to the Pig Stand throne and former president of Pig Stands, Inc. "As soon as they saw a Model T start to slow down, they'd race out to see who could jump up on the running board first, while the car was still moving."

An Explosion of Pork Barbecue

With its good food and derring-do curb service, the legend of the Texas carhop grew as the reputation of the Pig Stands and its signature sandwich spread. Propelled beyond the borders of Texas by franchising, the number of pork stands multiplied. Between 1921 and 1934, more than 100 Pig Stands were serving up "A Good Meal at Any Time" across America. Drive-in curb service had gone nationwide, and scores of operators duplicated the successful format.

In 1930, Royce Hailey, future patriarch of the Pig Stands clan and father to Richard, started as a Dallas carhop at age 13. Moving up through the ranks to take the president's job, he became sole owner in 1975. A self-made Texan with a knack for food, he's credited with inventing the chicken-fried steak sandwich and the super-sized slice of grilled bread called "Texas Toast." Food historians also cite onion rings as one of his more famous works of culinary art.

Modern Hard Times

Unfortunately, the novelty of the drive-in restaurant and the nostalgic comfort food it served wasn't enough to carry the operation into the new millennium. In recent years, all of the Texas locations have ceased car and dining room service for one financial reason or another. A single exception exists in San Antonio. Although it was closed with the others, what is now known as Mary Ann's Pig Stand was saved from the scrap heap of history when longtime employee Mary Ann Hill came up with the money to reopen it. Starting as a carhop at age 18 in 1967, Hill had never worked anywhere else. With its original Georges Claude neon pig-shaped sign, vintage Coke machine, shelves of pig memorabilia, and canopied lot, the restaurant operates under trustee status.

Fortunately, then, longtime fans and curious newcomers can still get a milkshake, a Pig Sandwich, and many of the classic fast food entrees that Hailey pioneered—including his signature Texas Toast and giant onion rings. For fans of "The World's First Drive-in Restaurant," there's still nothing that compares with chowing down in America's favorite dining room: the front seat of the car.

243

CHAPTER 9

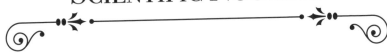

SCIENTIFIC NUGGETS

Rats! They're . . . Everywhere?

Urban folklore would have us believe that we're never farther than a few feet from a rat. The thought is enough to make your skin crawl, but are there really that many rats around us?

Why are rats so reviled? Not everyone hates rats. The Jainist religious sect of south Asia honors all life—even that of a rat. People love their pet rats. Your weird friend (you know which one) even likes wild rats. Of course, that might change when he contracts bubonic plague.

Beyond that, the only creatures that like rats are rat predators. The same animal lover who would feed and care for stray dogs would likely pay an exterminator good money to dispose of stray rats. Wild rats carry diseases and filth, eat unspeakable things, are very difficult to kill, can grow to an enormous size, and run in large packs that could overwhelm any human. To the majority of people, rats are the stuff of nightmares, as Winston Smith finds in George Orwell's *1984*.

So just how close to us are they? Do you spend a lot of time in the alleys of a large city's slums, cuddled up next to a garbage can, drinking in the smell of fermenting everything? Do you often seek shelter in the cool, tranquil comfort of your favorite sewer pipe? Do you spend idle afternoons sifting through that landfill you love so well in search of rare treasures?

If you answered yes to any of these questions, you've been in real close proximity to rats. Then again, if these are your preferred haunts, you know that already. For your safety, you might want to peruse the February 13, 1998, *Morbidity and Mortality Weekly Report*. In it you'll find an article that describes a couple of bona-fide cases of rat-bite fever.

Then we'll just avoid those places. Unfortunately, rats aren't picky about where they live. Some estimates say there is one rat per U.S. resident, which is hard to confirm because vermin don't answer the Census. But suppose there are that many. They'd be concentrated in big cities where there's also a lot of food and places to hide and scurry. Any poorly secured storage of food, either fresh or discarded, will attract them. People living in immaculate suburban mansions probably don't have a homey woodpile or trash heap in their backyards, but that's not to say rats don't roam idyllic family neighborhoods.

What do we do if we encounter a rat? The number of rats reported to health officials in the suburbs has been steadily increasing, and it's now common for municipalities to offer some sort of "rat patrol" to assist citizens in the fight against these critters. Have you heard the horror stories about rats that get into residential toilets after swimming up through sewer pipes? We'd like to say that those are also urban folklore—but they're not.

The Bear Truths

Although bearlike in appearance, with their rounded ears, plush fur, and black noses, koalas aren't actual bruins. Pandas, on the other hand, are true to form.

Cute as an Opossum?

There are few animals on Earth cuter than the cuddly koala. They're sometimes called Australia's teddy bears, but koalas are in fact related more closely to the ratlike American opossum than the impressive American grizzly.

Koalas are marsupials, which means they raise their young in special pouches, just like kangaroos, wallabies, and wombats. Their young, called joeys, are about the size of a large jelly bean when born and must make their way through their mother's fur to the protection of the pouch if they are to survive. As a baby grows, it starts making trips outside the pouch, clinging to its mother's stomach or back but returning to the pouch when scared, sleepy, or hungry. When a koala reaches a year old, it's usually large enough to live on its own.

Unlike real bears, koalas spend almost their entire lives roosting in trees, traveling on the ground only to find a new tree to call home. Koalas dine on eucalyptus leaves, of which there are more than 600 varieties in Australia. Eucalyptus leaves are poisonous to most other animals, but koalas have special bacteria in their stomachs that break down dangerous oils.

The Case for Pandas

Until recently, the giant panda was also considered a non-bear. Some scientists believed pandas were more closely related to the raccoon, whereas others speculated that they were in a group all their own. However, when they studied the animals' DNA, scientists were able to confirm that the giant panda is a closer relative to Yogi Bear than it is to Rocky Raccoon.

The Myth of the Big Easy

After Hurricane Katrina, lawmakers and media commentators questioned the wisdom of reconstructing New Orleans, a city that they claimed was "built ten feet below sea level."

According to geographical surveys, about 50 percent of New Orleans rests at or above sea level, and the other half rarely drops lower than minus six feet. Of the few points that lay 10 feet below sea level—approximately 50 acres of the 181 total square miles of the city—none have been developed for commercial use or habitation, serving instead as culverts, canals, and highway underpasses.

A Precarious Edge of Disaster

Geography, nature, and economics have conspired over the centuries to make New Orleans particularly vulnerable to floods. The location that became "the inevitable city on an impossible site" was chosen by French settlers in 1718 because it offered an easy portage between the Mississippi River— the main water artery for the interior of North America—

and Lake Ponchartrain, an outlet into the Gulf of Mexico for trading ships. (It was well into the nineteenth century before the mouth of the Mississippi became readily navigable for oceangoing ships.)

Isle d'Orleans

Although that land consisted mostly of cypress swamps, silt deposits left by seasonal floods over thousands of years created high ground along the banks of the river. The famous French Quarter—the original core of the city—was laid out on part of this elevated ground, some 15 feet above sea level. Over decades, the higher ground along the bending curve of the Mississippi was also settled, as the Crescent City grew behind an expanding system of levees. Because it was surrounded by a river, a lake, and swamps, the French referred to New Orleans as Isle d'Orleans, and for a century the city was a physical and cultural island, an outpost of the Caribbean on U.S. soil.

Developing the Land

The jewel of Jefferson's Louisiana Purchase, New Orleans grew faster than any other American city between 1810 and 1840, becoming the country's third largest after New York and Baltimore. But New Orleans's improbable geography limited its growth. The lower-lying land remained undeveloped until the early 1900s, when engineer A. Baldwin Wood invented an industrial-size screw pump for an ambitious drainage system that was built to protect the city from floods.

It's Sinking...

With nature seemingly under control, New Orleans expanded greatly during the twentieth century. The shores of Lake Pontchartrain, which represent some of the lowest-lying land, were filled in and new levees were built. But the levee system also stopped the seasonal deposits of silt that created most of the land in southern Louisiana. As a result, the city and much of its surrounding landscape is sinking as the loose soil settles without replenishment. Scientists estimate that the current rate of subsidence ranges from one-third to one and a half inches per year, with the possibility that 15,000 square miles will fall at or below sea level within 70 years.

Although the flooding of New Orleans after Hurricane Katrina was definitely a manmade disaster—the levee system designed by the U.S. Army Corps of Engineers failed to perform to its design specifications—the leveed "bowl" in which the city rests is inevitably subject to flooding.

Fallacies & Facts: Snakes

There are any number of myths about these fascinating creatures. Let's find out some of the truths.

Fallacy: You can identify poisonous snakes by their triangular heads.

Fact: Many non-poisonous snakes have triangular heads, and many poisonous snakes don't. You would not enjoy testing this theory on a coral snake, whose head is not triangular. It's not the kind of test you can retake if you flunk.

Boa constrictors and some water snakes have triangular heads, but aren't poisonous.

Fallacy: A coral snake is too little to cause much harm.

Fact: Coral snakes are indeed small and lack long viper fangs, but their mouths can open wider than you might imagine—wide enough to grab an ankle or wrist. If they get ahold of you, they can inject an extremely potent venom.

Fallacy: A snake will not cross a hemp rope.

Fact: Snakes couldn't care less about a rope or the material from which it's formed, and they will readily cross not only a rope but a live electrical wire.

Fallacy: Some snakes, including the common garter snake, protect their young by swallowing them temporarily in the face of danger.

Fact: The maternal instinct just isn't that strong for a mother snake. If a snake has another snake in its mouth, the former is the diner and the latter is dinner.

Fallacy: When threatened, a hoop snake will grab its tail with its mouth, form a "hoop" with its body, and roll away. In another version of the myth, the snake forms a hoop in order to chase prey and people!

Fact: There's actually no such thing as a hoop snake. But even if there were, unless the supposed snake were rolling itself downhill, it wouldn't necessarily go any faster than it would with its usual slither.

Fallacy: A snake must be coiled in order to strike.

Fact: A snake can strike at half its length from any stable position.

It can also swivel swiftly to bite anything that grabs it—even, on occasion, professional snake handlers. Anyone born with a "must grab snake" gene should consider the dangers.

Fallacy: Snakes do more harm than good.

Fact: How fond are you of rats and mice? Anyone who despises such varmints should love snakes, which dine on rodents and keep their numbers down.

Fallacy: Snakes travel in pairs to protect each other.

Fact: Most snakes are solitary except during breeding season, when (go figure) male snakes follow potential mates closely. Otherwise, snakes aren't particularly social and are clueless about the buddy system.

Fallacy: The puff adder can kill you with its venom-laced breath.

Fact: "Puff adder" refers to a number of snakes, from a common and dangerous African variety to the less aggressive hog-nosed snakes of North America. You can't defeat any of them with a breath mint, because they aren't in the habit of breathing on people, nor is their breath poisonous.

Pearls of Wisdom

Is it true that oysters produce pretty pearls? The answer isn't quite so black and white, round or smooth, say biologists.

Pearls do come from oysters, which are members of the mollusk family, but not all types of oysters produce pearls.

This is especially true of the kind we eat, so don't expect a pretty white surprise the next time you're enjoying oysters on the half shell at your favorite raw bar. Nor are oysters the only mollusks that produce pearls. Clams and mussels are also known to generate the coveted orbs.

Oysters are bivalves, which means their shell is made of two parts held together by an elastic ligament. An organ called a mantle manufactures the oyster's shell, which is lined with a substance called nacre—a beautiful iridescent material made of minerals derived from the oyster's food.  Pearls are produced when a foreign substance, such as a grain of sand, becomes embedded in the shell and irritates the mantle. To reduce the irritation, the mantle covers the substance with layer upon layer of nacre. Over time, this creates a pearl.

Natural pearls are most prized, but because of the way they're produced, they're not the most common variety. To meet consumer demand, most pearls used in jewelry are cultivated—which involves making a tiny cut in an oyster's mantle and inserting an irritant. The oyster does the rest of the work.

Another common myth about oysters is that it's safe to eat them only in months that contain the letter "r"—September through April. It is thought that oysters harvested in the remaining four months are more likely to carry harmful bacteria, which could cause food poisoning. According to the Centers for Disease Control and Prevention, however, oyster-related bacterial illnesses occur year-round.

Cocky Cockroaches

Would the cockroach be the sole survivor of a nuclear war? No, but the mightiest of the mighty still looks up to this creepiest of crawlers.

We've all heard the scenario: A swarm of nuclear missiles is launched in unison, aimed at strategic targets throughout the world. The end of humankind is assured. In the span of a few minutes, entire civilizations are obliterated and the vital peoples and creatures that once roamed the planet are left dead or dying. Soon all life forms will succumb to the insidious effects of radiation. All except one, that is.

Blattodea, better known as the common cockroach, is the tough-guy holdout in this nightmarish scene. But would it be?

The pesky insect might be forgiven for flaring its chest with pride. It has the ability to withstand doses of radiation that would easily kill a human, It is generally accepted that a human will perish after receiving a 400- to 1,000-rad dose of radiation. In contrast, the hardy cockroach can withstand a 6,400-rad hit and continue crawling, which suggests that this bothersome bug is champ.

But the cockroach shouldn't gloat too much. It may outlast a human, but the insect kingdom boasts tougher players still in the form of tiny fruit flies. Fruit flies may be small, but they sure are resilient. The Caribbean fruit fly, the "mighty mite" among these little fellows, takes a 180,000-rad dose like it's nothing. This figure beats the roach by a factor of ten.

Stick Around

What happens if you swallow a stick of chewing gum? Will it stick around for seven years? Twist around your innards? Form a blockage in your digestive tract?

It's called chewing gum, not swallowing gum. But sometimes, accidentally or on purpose, a piece of gum ends up dropping down the gullet. When that happens, who hasn't wondered what the consequences will be?

No one knows how it got started, but the idea that swallowed chewing gum stays in the digestive system for seven years is a pervasive myth. It seems the misconception dates back thousands of years, as archaeologists have found evidence of ancient wads of chewing gum. Way back when, gum didn't come wrapped in paper and foil, but the concept was the same—it was something pleasant to chew on but not to swallow.

No matter how old the myth, you need not worry about swallowed gum taking up long-term residence in your stomach. Gastroenterologists say that inspections of the digestive tract, with exams such as colonoscopies and endoscopies, do not reveal clumps of petrified gum. When gum does show up on such scans, it is most often a recent arrival.

Although it's not intended to be ingested, chewing gum usually is not harmful if it ends up in your stomach instead of a trash bin (or under a desk). Some chewing gum additives, such as sweeteners and flavoring, are broken down by the body, but the bulk of gum is not digestible. Ingredients such as rubbery elastomers and resins remain intact during their slow voyage through the digestive tract. Eventually, the gum moves down and out.

In rare cases, an extremely large clump of swallowed gum could get stuck on its journey out of the body, causing a dangerous blockage. This potential problem can be avoided, however, if you chew just one stick at a time.

A Question of Canine Cleanliness

Smooch your pooch at your own risk. It's not man's best friend that could kiss and kill.

Most dog owners will tell you that their dog's mouth is much cleaner than a human's. In fact, this old wives' tale has been touted so loudly and for so long that most people assume it's true. Most veterinarians, however, disagree. They'll tell you it's a stalemate—both human and canine mouths are rife with bacteria.

One of the biggest reasons people believe the myth is the fact that dogs lick their wounds, and those wounds tend to heal very quickly. But it's not as though a dog's saliva has amazing antibacterial properties. Dogs' cuts and scrapes get better fast because their tongues help get rid of dead tissue and stimulate circulation, which in turn facilitates the healing process.

You Know Where It's Been

If you still think a dog's tongue is more antiseptic than your own, just take a look at what your pet's tongue touches over the course of a day. Dogs use their tongues for eating and drinking, as well as for activities such as bathing and exploring garbage cans and weird dead things in the yard.

A dog bite, like a human bite, can cause infection if it breaks the skin. But the bacteria transmitted in each are fairly species-specific. In other words, a bug that's harmful to humans likely won't be transmitted to your pooch if you give him a big, slobbery kiss on the mouth, and vice versa.

At least your dog's bite isn't as dangerous as that of a Komodo dragon. The mouth of this giant, carnivorous lizard, indigenous to Indonesia, is a veritable petri dish of disgusting bacteria, many of which can cause an agonizing, often fatal infection.

A Quicksand Sham

If you happen to get trapped in one of nature's suction pits, hang loose. Your moves are the only things that can drag you down.

Let's say you're running through the woods and you trip and fall. As you attempt to right yourself, you realize that the earth below you isn't really earth at all, and you find it impossible to find purchase. You are wet and covered in a granular grime, but it's not a body of water or sand pit that you've fallen into. This substance seems more like a combination of the two. In fact, it is—you have fallen into a quicksand pit. What you do from this point forth will determine whether this will be a momentary inconvenience or just a slightly longer inconvenience.

A Quicksand Primer

Quicksand is a sand, silt, or clay pit that has become hydrated, which reduces its viscosity. Therefore, when a person is "sucked" down, they are in reality simply sinking as they would in any body of water.

So why does quicksand make people so nervous? It's probably due to the fact that it can present resistance to the person who steps in it. This is particularly true of someone who is wearing heavy boots or is laden with a backpack or other load. Obviously, the additional weight will reduce buoyancy and drag one down.

It's All in the Legs

In the human thirst for drama, quicksand has a reputation as a deadly substance. The facts show something different. Because quicksand is denser than water, it allows for easy floating. If you stumble into a pit, you will sink only up to your chest or shoulders. If you want to escape, all you need to do is move your legs slowly. This action will create a space through which water will flow, thereby loosening the sand's grip. You should then be able to float on your back until help arrives.

There's More to Know About Tycho

A golden nose, a dwarf, a pet elk, drunken revelry, and . . . astronomy? Read about the wild life of this groundbreaking astronomer.

Look to the Stars

Tycho Brahe was a Dutch nobleman who is best remembered for blazing a trail in astronomy in an era before the invention of the telescope. Through tireless observation and study, Brahe became one of the first astronomers to fully understand the exact motions of the planets, thereby laying the groundwork for future generations of star gazers.

In 1560, Brahe, then a 13-year-old law student, witnessed a partial eclipse of the sun. He reportedly was so moved by the event that he bought a set of astronomical tools and a copy of Ptolemy's legendary astronomical treatise, Almagest, and began a life-long career studying the stars. Where Brahe would differ from his forbears in this field of study was that he believed that new discoveries in the field of astronomy could be made, not by guesswork and conjecture, but rather by rigorous and repetitious studies. His work would include many publications and even the discovery of a supernova now known as SN 1572.

Hven, Sweet Hven

As his career as an astronomer blossomed, Brahe became one of the most widely renowned astronomers in all of Europe. In fact, he was so acclaimed that when King Frederick II of Denmark heard of Brahe's plans to move to the Swiss city of Basle, the King offered him his own island, Hven, located in the Danish Sound.

Once there, Brahe built his own observatory known as Uraniborg and ruled the island as if it were his own personal kingdom. This meant that his tenants were often forced to sup-ply their ruler (in this case Brahe) with goods and services or be locked up in the island's prison. At one point, Brahe imprisoned an entire family—contrary to Danish law.

Did We Mention That He Was Completely Nutty?

While he is famous for his work in astrono-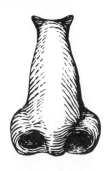
my, Brahe is more infamous for his colorful
lifestyle. At age 20, he lost part of his nose
in an alcohol-fueled duel (reportedly using
rapiers while in the dark) that ensued after
a Christmas party. Portraits of Brahe show
him wearing a replacement nose possibly
made of gold and silver and held in place
by an adhesive. Upon the exhumation of
his body in 1901, green rings discovered around the nasal cavity of Brahe's skull have also led some scholars to speculate that
the nose may actually have been made of copper.

While there was a considerable amount of groundbreaking
astronomical research done on Hven, Brahe also spent his time
hosting legendarily drunken parties. Such parties often featured a colorful cast of characters, including someone named
Jepp who dwelled under Brahe's dining table and functioned
as something of a court jester; it is speculated that Brahe
believed that Jepp was clairvoyant. Brahe also kept a tame pet
elk, which stumbled to its death after falling down a flight of
stairs—the animal had gotten drunk on beer at the home of a
nobleman.

Brahe also garnered additional notoriety for marrying a
woman from the lower classes. Such a union was considered
shameful for a nobleman such as Brahe, and he was ostracized
because of the marriage. Thusly all of his eight children were
considered illegitimate.

However, the most lurid story of all is the legend that Brahe died
from a complication to his bladder caused by not urinating,

out of politeness, at a friend's dinner party where prodigious amounts of wine were consumed.

The tale lives on, but it should be pointed out that recent research suggests this version of Brahe's demise could be apocryphal: He may have died of mercury poisoning from his own fake nose.

If These Bones Could Talk

Early in the twentieth century, archaeologists searched frantically for the "missing link"—a fossil that would bridge the gap between apes and man. What was found, however, made monkeys out of everyone involved.

Fossil Facts or Fiction?

In November 1912, a story appeared in the English newspaper per *Manchester Guardian*: Skull fragments had been found that could be of the utmost significance. "There seems to be no doubt whatever of its genuineness," wrote the reporter, characterizing the bones as perhaps "the oldest remnant of a human frame yet discovered on this planet." The story generated feverish speculation. On the night of December 18, 1912, a crowd jammed into the meeting of the Geological Society of London to learn about this amazing discovery.

What they heard was that solicitor and amateur archeologist Charles Dawson had discovered two skull fragments and a jawbone from a gravel bed near Piltdown Common in East Sussex. He had been interested in this area ever since workmen, knowing of his archeological interest, had given him some interesting bone fragments from the pit several years before. Dawson had since been making his own excavations of the pit, aided by Arthur Smith Woodward, keeper of the Department of Geology at the British Museum.

The skull fragments were definitely human, but the jawbone was similar to an ape. If they came from the same creature, as Woodward and Dawson both hypothesized, then the two men had discovered the missing evolutionary link between ape and human. Woodward announced, "I therefore propose that the Piltdown specimen be regarded as a new type of genus of the family Hominidae."

A Deep Divide

Almost immediately, two distinct camps were formed: doubters and supporters. In Woodward's favor were the facts that the remains were found close together, that they were similar in color and mineralization, and that the teeth were worn down in a flat, human way—unlike those of an ape. Doubters contended the jawbone and skull fragments were too dissimilar to be from the same creature. American and French scientists tended to be skeptical, while the British generally accepted the validity of the discovery.

Woodward's side scored valuable points when a canine tooth missing from the Piltdown jaw was discovered in 1913 close to where the jawbone originally had been found. Hard on the heels of that find came another—an elephant bone that had been rendered into some type of tool and supposed to have been used by Piltdown Man.

In 1915, there came perhaps the most conclusive evidence of all: Dawson found the remains of a similar creature a scant two miles away from the site of the first discovery.

Bone Betrayal

So, Piltdown Man entered the ar-chaeological record. After Dawson died on August 10, 1916, no signifi-cant new Piltdown discoveries were made, but no matter. Even when a few scientists identified the jaw as that from an ape, they were ignored.

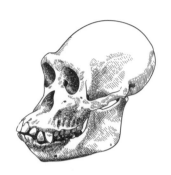

However, as other fossil discoveries were made in subsequent years, it became evident that something wasn't quite right about Piltdown Man. Things began unraveling in 1949, when a new dating technique called the fluorine absorption test was used on Piltdown Man. The test revealed that both the skull fragments and the jawbone were relatively modern. Finally, in 1953 a group of scientists proved conclusively that Piltdown Man was a hoax. The jawbone had been stained to look old, the teeth filed down, and the bones placed at the site.

Although the identity of the Piltdown Man hoaxer has never been revealed—even Sir Arthur Conan Doyle, author of the *Sherlock Holmes* series of mysteries, is considered a suspect by some—most suspicion falls on Dawson, who was later found to have been involved in other archeological frauds. Ultimately, it seems that if seeing is believing, then Piltdown Man is proof that people will only see what they want to believe.

The Mysterious Blue Hole

State Route 269 hides a roadside attraction of dubious depth and mysterious origin, a supposedly bottomless pool of water that locals simply call the "Blue Hole."

Every state has its tourist traps and bizarre little roadside attractions that are just intriguing enough to pull the car over to see. Back in the day, no roadside attraction brought in the Ohio travelers more than a bottomless pond filled with blue water: the mysterious Blue Hole of Castalia.

The Blue Hole's Origins

The Blue Hole is believed to have formed around 1820, when a dam burst and spilled water into a nearby hole. The ground surrounding Castalia is filled with limestone, which does not absorb groundwater well. The water quickly erodes the limestone, forming cave-ins and sinkholes. It wouldn't be until the late 1870s, however, that most people were made aware of the Blue Hole's existence; the hole was in a very isolated location in the woods. Once the Cold Creek Trout Club opened up nearby, however, its members began taking boat trips out to see the hole, and people all over the area were talking about the mysterious Blue Hole hiding out in Castalia. In 1914, a cave-in resulted in the Blue Hole growing to its current size of almost 75 feet in diameter.

Stop and See the Mystery

The owners of the property where the Blue Hole is situated began promoting it as a tourist stop beginning in the 1920s. It didn't hurt that the entrance to the Blue Hole property was along State Route 269, the same road that people took to get to Cedar Point amusement park. It is estimated that, at the height of its popularity, close to 165,000 people a year came out to take a peek at the Blue Hole.

The Blue Hole was promoted as being bottomless. Other strange stories were often played up as well, including the fact that the water temperature remained at 48 degrees Fahrenheit year-round. Tour guides would point out that regardless of periods of extreme rainfall or even droughtlike conditions, the Blue Hole's water level remained the same throughout.

So, What's Up with This Hole, Anyway?

Despite the outlandish claims and theories surrounding the Blue Hole and its origins, the facts themselves are rather mundane. The Blue Hole is really nothing more than a freshwater pond. It isn't even bottomless. Sure, the bright blue surface of the water does indeed make the hole appear infinitely deep, but in fact, it's really only about 45 feet to the bottom even at its deepest parts.

The blue color of the water is from an extremely high concentration of several elements, including lime, iron, and magnesium. That's the main reason there are no fish in the Blue Hole; they just can't survive with all that stuff in the water.

One Hole or Two?

During the 1990s, the owners of the Blue Hole fell on hard times, forcing them to close the attraction. Families who would show up at the front entrance were forced to stare sadly through a locked gate at the small trail into the woods.

That is until several years ago, when the nearby Castalia State Fish Hatchery began clearing land to expand its hatchery. Lo and behold, workers uncovered a second Blue Hole.

Just how this second Blue Hole came to be is still unknown, although the popular belief is that both holes are fed by the same underground water supply. None of that seems to matter to the Blue Hole faithful—they're just thankful to be able to take a gander at a Blue Hole again.

The Cincinnati Observatory

In 1842, Ormsby M. Mitchel started a series of public astronomy lectures that fascinated the Ohio River city. He asked people to donate money—as shareholders—to establish the Cincinnati Astronomical Society. Its mission: to buy a well-crafted telescope and build an observatory to house it. To Mitchel's surprise, 300 people pledged money.

It was in 1842 that Ormsby M. Mitchel traveled to Munich to inspect a 12-inch objective lens manufactured by a German optical factory. He ordered one and returned home to start constructing the Cincinnati Observatory on four acres donated by wealthy businessman Nicholas Longworth. It stood atop Mount Ida, 400 feet above the growing city.

On November 9, 1843, former President John Quincy Adams participated in laying the observatory's cornerstone.

He delivered his last public speech that day. A believer in the sciences, Adams had tried, unsuccessfully, to persuade Congress to build a national observatory. When he left town, the appreciative city renamed the hill in his honor—Mount Adams.

By the time the telescope arrived from Germany in January 1845, the nation had slipped into an economic depression. Because most of the pledged money had been spent to buy the telescope, Mitchel had to use his own money and raise more from the public to start constructing the building. Many workers agreed to help at no charge.

On April 14 of that year, Cincinnati's telescope started operating. Mitchel discovered a companion to the bright star Antares. He also founded *The Sidereal Messenger*, the first astronomical publication in the United States. Just as the observatory seemed ready for great work, the Cincinnati College burned, leaving Mitchel without an employer and the observatory with no source of funds. He agreed to serve as director without pay while the society limped along on donations.

Enduring and Reopening

Still without a paycheck by the early 1850s, Mitchel moved to Albany, New York, to help develop a new observatory. When the Civil War started in 1861, Mitchel returned to Ohio to become a Union general. After helping Cincinnati build defensive positions, he moved on to other missions. He died of yellow fever while serving in South Carolina in 1862.

The Cincinnati Observatory, meanwhile, was closed during the war. In 1868 it reopened, and a new director, Cleveland Abbe, moved it to neighboring Mount Lookout to avoid the city's heat, dust, and smoke. The new observatory opened in 1873, using the old telescope. But Abbe accomplished more than scanning the skies: He established regular weather reports and storm predictions, for which he was nicknamed "Old Probabilities." Impressed by his predictions, the federal government hired him to come to Washington, D.C., to establish the United States Weather Bureau.

Again, the Cincinnati Observatory closed temporarily. But in time, it hired new directors and expanded. In 1904, a 16-inch refractor telescope was installed in the old building, and a new building was constructed to house the original reflector telescope, with a shortened tube.

In 1997, preservation became a priority when local astronomy enthusiasts formed the Cincinnati Observatory Center as a volunteer committee that was dedicated to saving the observatory in its historical surroundings. Late that same year, the group won designation for the observatory as a National Historic Landmark from the U.S. Department of the Interior.

Today, the nation's oldest observatory serves more than 6,000 people annually, both youths and adults. "There are a lot of closet astronomers out there," said Dean Regas, staff outreach astronomer. "We appeal to them and to a lot of other people who just want to look at the stars."

Do People Really Go Mad During a Full Moon?

The story goes that if you ask emergency room workers or police officers, they'll tell you that the number of disturbed individuals who come to their attention rises dramatically during a full moon. What's the truth behind the story?

It's a long-held belief that a lunar effect causes "lunacy" in susceptible people—resulting in an increase in homicides, traffic accidents, suicides, kidnappings, crisis calls to emergency services, admissions to psychiatric institutions, and all kinds of other things. The rationale: The earth is 80 percent water, and so is the human body. Theoretically, then, since the moon has such a dramatic effect on the tides, it could move the water in our bodies in some similar way, causing strange behavior.

Are the stories true? Most evidence says no.

In 1996, scientists Ivan Kelly, James Rotton, and Roger Culver did a thorough examination of more than one hundred studies of lunar effects. Perhaps surprisingly, they found no significant correlation between the state of the moon and people's mental and physical conditions. When all of the statistical wrinkles had been smoothed out, there was no evidence of a rise in violence, accidents, disasters, or any other kind of strange behavior.

A study by C. E. Climent and R. Plutchik, written for *Comprehensive Psychiatry*, showed that psychiatric admissions are lowest during a full moon, and an examination conducted at the University of Erlangen–Nuremberg in Germany indicated no connection between suicide rates and phases of the moon. So why the myths? Perhaps people just want to believe the spooky tales, and lunar effects are tossed into movies and literature simply because they're compelling drama. Who doesn't love a good werewolf tale?

The constant reinforcement of the "full moon" message makes it much more likely that the public will accept it as proven fact.

Myths also tend to stay alive if you pick and choose the data to fit the story. One murder that occurs during a full moon creates a story that can be told over and over, yet the 10 homicides that happen at any other time of the month just disappear into a pile of statistics.

Renowned UCLA astronomer George O. Abell consistently dismissed claims that the moon could have a strong enough effect on the water in a human body to cause any behavioral changes. Abell pointed out that a mosquito would exert more gravitational pull on a human arm than the moon ever might.

So, is that arm-biting mosquito spooked by the moon? That's not entirely a joke, because it seems animals actually are affected by lunar activity. It might be a bit scary to read a study in the *British Medical Journal* that appears to prove there's a significant increase in bites by cats, rats, horses, and dogs when the moon is full.

Suddenly, the image of a dog howling at the moon might give you a little shiver. But a man baying at the moon? Most likely he'd be someone who's goofy all month long.

If Bats Are Blind, How Do They Know Where They're Going?

Despite the phrase, bats can see. Some have pretty good vision.

Take the megachiroptera bat (more commonly known as the Old World fruit bat or flying fox). Members of this tropical suborder are known for their large eyes and excellent nighttime eyesight. Studies have shown that they're able to see things at lower light levels than even humans can. Most Megachiroptera bats rely completely on their vision to find the fruits and flower nectar they like to munch.

Smaller Microchiroptera bats count on their eyesight, too. These insect-eating bats can see obstacles and motion while navigating speedy, long-distance trips. However, like many bat species, mouselike Microchiroptera also receive some extra guidance from a remarkable physiological process known as echolocation. When flying in the dark, these bats emit high-frequency sounds and then use the echoes to determine distance and direction, as well as the size and movement of anything in front of them.

This "biological sonar system" is so refined that it can track the wing beats of a moth or something as fine as a human hair. Neuroethologists (people who study how nervous systems generate natural animal behavior) will tell you that our military doesn't even have sonar that sophisticated.

Bats have been the subject of myth, mystery, and misconception for centuries. Until recently, traditional thinking was that nocturnal bats could see at night but were blind by day. Now scientists at the Max Planck Institute for Brain Research in Frankfurt, Germany, and at The Field Museum of Natural History in Chicago have discovered that Megachiroptera bats have daylight vision, too. Apparently, this vision comes in handy for locating predators and even for socializing. Flying foxes don't sleep all day—they bounce from treetop to treetop for daytime confabs with their batty neighbors.

So, there you have it: Bats can see, and they know where they are going. The next time you want to use a creative—though rather impolite—idiom to describe nearsightedness, you'd be more accurate to say "blind as a mole." That small, burrowing mammal has very small eyes and, indeed, very poor vision.

Teleportation: Not Just the Stuff of Science Fiction

Scientists say that it's only a matter of time before we're teleporting just like everyone does on *Star Trek*.

Beam Us Up

We're closer than you might think to being able to teleport, but don't squander those frequent-flyer miles just yet. There's a reason why Captain Kirk is on TV late at night shilling for a cheap-airfare Web site and not hawking BeamMeToHawaiiScotty.com. For the foreseeable future, jet travel is still the way to go.

If, however, you're a photon and need to travel a few feet in a big hurry, teleportation is a viable option. Photons are sub-atomic particles that make up beams of light. In 2002, physicists at the Australian National University were able to disassemble a beam of laser light at the subatomic level and make it reappear about three feet away. There have been advances since, including an experiment in which Austrian researchers teleported a beam of light across the Danube River in Vienna via a fiber-optic cable—the first instance of teleportation taking place outside of a laboratory.

These experiments are a far cry from dematerializing on your spaceship and materializing on the surface of a strange planet to make out with an alien who, despite her blue skin, is still pretty hot. But this research demonstrates that it is possible to transport matter in a way that bypasses space—just don't expect teleportation of significant amounts of matter to happen until scientists clear a long list of hurdles, which will take many more years.

Here, Gone, There

Teleportation essentially scans and dematerializes an object, turning its subatomic particles into data. The data is transferred to another location and used to recreate the object. This is not unlike the way your computer downloads a file from another computer miles away. But your body consists of trillions upon trillions of atoms, and no computer today could be relied on to crunch numbers powerfully enough to transport and precisely recreate you elsewhere.

As is the case with many technological advances, the most vexing and long-lasting obstacle probably won't involve creation of the technology, but rather the moral and ethical issues surrounding its use. Teleportation destroys an object and recreates a facsimile somewhere else. If that object is a person, does the destruction constitute murder? And if you believe that a person has a soul, is teleportation capable of recreating a person's soul within the physical body it recreates? These are questions with no easy answers.